RELATIVITY PHYSICS

Student Physics Series

RELATIVITY PHYSICS

Roy E. Turner

Reader in Theoretical Physics
University of Sussex

ROUTLEDGE & KEGAN PAUL
London, Boston, Melbourne and Henley

First published in 1984
by Routledge & Kegan Paul plc
39 Store Street, London WC1E 7DD, England
9 Park Street, Boston, Mass. 02108, USA
464 St Kilda Road, Melbourne,
Victoria 3004, Australia and
Broadway House, Newtown Road,
Henley-on-Thames, Oxon RG9 1EN, England

Set in Press Roman by Hope Services, Abingdon
and printed in Great Britain
by Cox & Wyman Ltd., Reading

Library of Congress Cataloging in Publication Data

Turner, R. E. (Roy Edgar), 1931–

Relativity physics.
(Student physics series)
Bibliography: p.
Includes index.
1. Relativity (Physics) I. Title. II. Series.
QC173.55.T87 1984 503.1'1 83-24707

ISBN 0-7102-0001-3

To my wife Johanna, who being
a more experienced author than
I am, has sympathetically
understood, if not the subject
matter of this book, the
problems of its production

Contents

Preface

It is somewhat presumptuous to produce yet another book on relativity, despite the fact that the publishers are producing a new series of texts on undergraduate physics. There is already a number of excellent texts on the market which deal clearly with many aspects of the subject. As always in this situation, the author is reduced to justifying his efforts by claiming that he hopes he has produced a new perspective on the subject. In my case that perspective is not really a new one but goes back to the arguments that were around at the beginning of this century. Most texts at undergraduate level omit, or treat very scantily, the role of electromagnetic theory in the development of special relativity, and yet it played a central role.

I have tried in this book to present mechanics and electromagnetism in the context of special relativity, to explain, in turn, the problems that arose when the principle of relativity was applied to each of them and how Einstein showed that, once the leap in imagination was made, the solution was relatively simple. It is an exciting story and I hope that the book conveys, if only in parts, that excitement. But above all I hope that undergraduates will enjoy reading it, for physics is an enjoyable subject to try to understand.

The reader should have had a good grounding in Newton's equations of motion and an understanding of electromagnetism up to Maxwell's equations and their plane-wave solutions. Quantum mechanics is mentioned but the reader is only required to know the Planck and de Broglie relations. Some readers may find Chapter 7 hard going. The chapter is there for three reasons.

First, it brings together the two subjects that dominate this book, mechanics and electromagnetism. Second, relativistic equations of motion can be written down (despite the subsequent difficulties that they lead to) and very rarely are. Finally I hope it will encourage the interested student to go on and pursue the subject further.

Notation is very important in producing a clear presentation. I have steered away from tensor notation because, elegant though the equations look in this form, I believe that in a first reading it obscures the physics. Ordinary three-vectors are denoted as usual in bold type and their magnitude in normal type. Four-vectors are the same except they are distinguished by an arrow above them. Thus for example, the three-momentum is denoted by $\mathbf{p}$ and its magnitude by p. The four-momentum or energy-momentum four-vector by $\vec{\mathbf{p}}$ and its 'magnitude' by $\vec{p}$.

Finally I should like to thank my colleagues in the physics division for the many questions that I have put to them over the years that I have given a course on this subject at the University of Sussex. This particularly applies to Drs G. Barton, J. Byrne and J. Plaskett who have tolerated my pestering with good humour. Any mistakes are of course my own.

Chapter 1
The principle of relativity

1.1 Introduction

There are very few discoveries that can be said, unequivocally, to have changed radically the course of human progress. In evaluating the importance of the subject of this book it is possible to go even further and say that the formulation of special relativity, at the beginning of this century, was the most significant event in the history of mankind. The reasons for this are seen in the simple, and probably the most publicized, scientific formula

$$E = mc^2 \quad [1.1]$$

which shows the equivalence of energy, E, and mass, m, their values being related by the square of the velocity of light, c. From this apparently innocent-looking result, modern technology has developed the seeds of mankind's own destruction, a fact that was graphically and cruelly illustrated at Hiroshima and Nagasaki. The enormous amount of energy that is released in the atomic bomb is due to the factor c^2, since c is a very large number, $3 \times 10^8\,\mathrm{ms}^{-1}$. Thus a mass of 1 kg is equivalent to 9×10^{16} J of energy. (Compare this with the annual output of a 500 MW power station, a typical size for the UK which is equal to $5 \times 10^8 \times (60)^2 \times 24 \times 365 = 1.6 \times 10^{16}$ J.)

Although equation [1.1] is the most quoted result from special relativity, its derivation by Einstein in 1905 was not the most surprising or the most contentious result for the then contemporary physicists. As early as 1881 Sir J. J. Thomson considered a model of the electron which consisted of a uniform charge distributed over the surface of a sphere. The electrostatic energy of the electron is then

$$E = e^2/8\pi\epsilon_0 a$$

where e is the total electron charge and a is the radius of the sphere. He further observed that for a charge e moving with a uniform velocity $\boldsymbol{v}$ the energy of its field is $m_{\text{elm}} \boldsymbol{v}^2/2$, where the electromagnetic 'mass' is $m_{\text{elm}} = e^2/6\pi\epsilon_0 c^2 a$. Combining this result with the expression for the energy gives

$$E = \frac{3}{4} m_{\text{elm}} c^2$$

(For a detailed discussion of these results see Feynman, vol. II. For a description of other theories linking energy and mass see Whittaker, vol. 2.)

Another feature of the Einstein theory, namely that the length of an object is not the same when measured by a stationary observer as when measured by a moving observer, was also foreshadowed by Lorentz and FitzGerald. This phenomenon is still known as the Lorentz-FitzGerald contraction.

Despite these precursors, Einstein's theory provoked considerable controversy when it was published. The reasons for this will, I hope, emerge in the course of reading this book. The presentation given here will not be an historical one for three reasons. First, no physical theory proceeds smoothly from one development to the next – many blind and unprofitable alleys are followed. Second, many of the significant historical landmarks are lost in the mists of time. For example, whether Einstein knew of a critical experiment – the Michelson-Morley experiment – is a matter of controversy. (This controversy may have been resolved; see *Physics Today*, vol. 35, no. 8, Aug. 1982.) Thirdly, an excellent and highly readable account is given in Whittaker's classic work *A History of the Theories of Aether and Electricity* (see the bibliography). Despite forgoing the strictly historical approach, however, some of the original chronology will inevitably be followed, not only because this is the logical description but also because, hopefully, some of the original excitement will be conveyed.

As we shall see, special relativity is not a theory in the same sense as solid state physics or electromagnetism. What relativity

—or, more precisely, the principle of relativity—is, is a kind of 'check' on the validity of other theories. Crudely speaking the principle of relativity says that the laws of physics are the same for a stationary observer as they are for an observer moving with a constant velocity with respect to the first. Thus we can apply the principle of relativity to any physical theory and see if it is satisfied. If it is not, then we assert that the theory cannot be wholly correct but, at the best, can only be an approximation to a correct, relativistically invariant theory. (The precise meaning of the phrase 'relativistically invariant' will be given later. Here it may be taken to mean 'satisfies the principle of relativity'.) In fact for over two centuries it was believed that Newton's equations gave a correct description of the motion of material bodies. Following Einstein's discoveries, however, they were found not to satisfy the principle of relativity. Given their experimental verification, the conclusion to be drawn (and this also followed from the mathematics) was that in some sense Newton's equations were only valid as an approximation. The approximation was that the velocities of all the particles should be small compared with the velocity of light, a condition well satisfied by the motion of the planets. The Earth's velocity around the Sun, for example, is approximately $3 \times 10^4 \, ms^{-1}$ and the velocity of light is $3 \times 10^8 \, ms^{-1}$, so Newton's solution for the motion of the Earth should be accurate to one part in 10^4. (In fact it is the square of the velocity ratio which is important, so that the limit of validity is even better—one part in 10^8.)

Finally in this introduction a comment on the use of the word 'relativistic'. We are using it combined with the word 'invariant' to mean a theory which satisfies the principle of relativity. It is common these days, however, to refer to 'relativistic expressions' for some physical quantity, e.g. momentum. What this really means is that the expression so referred to is consistent with the principle of relativity. This is a little misleading, since Newton's expression for momentum, for example, is approximately compatible (see the discussion above) with the principle of relativity but is not described as relativistic. Probably the more correct way to describe the expressions derived by Einstein and others is 'fully relativistic', as opposed to the Newtonian expressions which are only 'approximately relativistic'. We shall try to avoid the adjective 'relativistic'.

This book is about a number of branches of physics (mainly two) in which the correct application of the principle of relativity (as realized by Einstein) radically altered their development. It may thus be reasonably labelled 'Relativity Physics'.

1.2 Reference systems

Physics is about measurement and the relationship of one measurement to another. Even in the most primitive societies there are two concepts which are intuitively obvious and present no difficulty in understanding. These two concepts are space and time. Furthermore, these two concepts are uncluttered, not being endowed with mystical or magical qualities. Very early on in his development man put these concepts on a quantitative basis. He used the rotation of the Earth as his clock and introduced an arbitrary length against which all other lengths were to be compared. For each of these measurements he chose a unit. For time the unit eventually became the second, defined until relatively recently (the modern definition will be given later) as $(1/24) \times (1/60) \times (1/60)$ of the period of rotation of the Earth. For distance the arbitrary length has varied from culture to culture, e.g. the idyllic-sounding rod, pole or perch and the rather dull-sounding metre.

If we are concerned with measuring distance—for example, the distance travelled by a particle—then this is most easily done by setting up a coordinate system or reference frame. The simplest such reference frame is a Cartesian one, which consists of three axes at right angles, labelled x, y and z. The position of any event is then given by specifying the coordinates (x,y,z), i.e. the distance one has to move along the x, y and z axes to the point in space where the event occurred. Figure 1.1 shows the trajectory of a particle and (x_1, y_1, z_1) and (x_2, y_2, z_2) are the coordinates of two events, namely the particle passing two markers.

For the measurement of time we need a clock which, with reference to some arbitrary zero, will specify how many seconds have elapsed before a particular event occurs. The instrument we

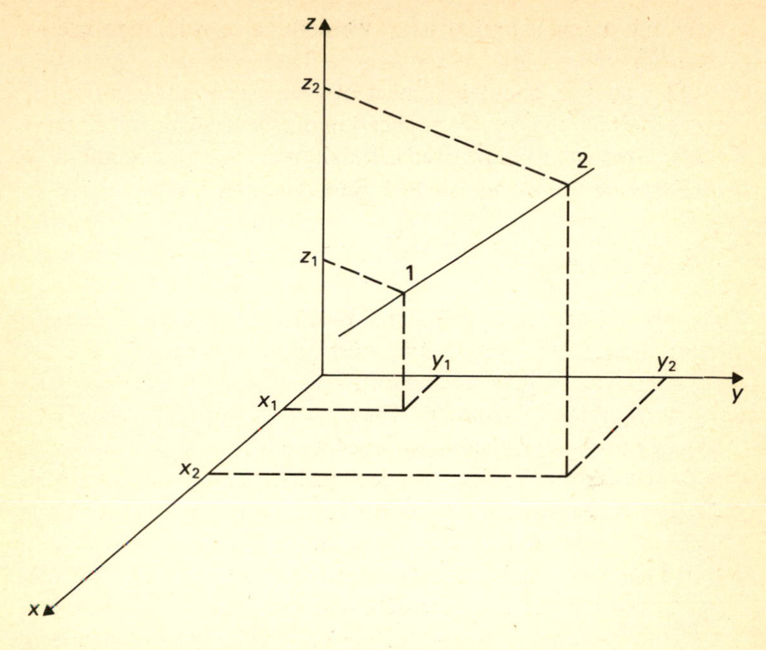

Fig. 1.1 The trajectory (linear) of a particle that passes two markers with coordinates (x_1, y_1, z_1) and (x_2, y_2, z_2).

use as a clock will be discussed in Chapter 3, but for the moment a laboratory stopclock will suffice. Thus for the two events illustrated in Fig. 1.1 we can add a fourth 'coordinate', time. The particle passes the first marker at a position in space given by (x_1, y_1, z_1) and does so at time t_1. We say it has the space–time coordinates (x_1, y_1, z_1, t_1). Similarly for event 2 we have the space–time coordinates (x_2, y_2, z_2, t_2). Thus to specify the coordinates of any event completely we need to know the three spatial components and the time; if we use a vector notation, this may be shortened to $(\mathbf{r}, t)$, where

$$\mathbf{r} = \hat{\mathbf{i}}x + \hat{\mathbf{j}}y + \hat{\mathbf{k}}z \qquad [1.2]$$

is the position vector.

Now the absolute values of $(\mathbf{r}, t)$ cannot be important because where we have the origin of the coordinate system is quite arbitrary, as indeed was the choice of the zero of time. In fact we are only interested in relative positions. Thus we might wish to know the time between the two events depicted in Fig. 1.1, i.e.

$$\Delta t = t_2 - t_1$$

This is independent of the origin of time. This is because if we shift the origin of time by an arbitrary amount, say t_0, then the new absolute times are $t_1' = t_1 - t_0$ and $t_2' = t_2 - t_0$, but the difference between the two, as we can see, remains the same. Similarly the distance travelled by the particle is independent of the origin of the reference frame. From Fig. 1.1 it is clear by simple geometry that

$$l^2 = (x_2 - x_1)^2 + (y_2 - y_1)^2 + (z_2 - z_1)^2 \qquad [1.3]$$

where l is the distance travelled by the particle. Since this expression depends only upon the differences of the coordinates, by arguments similar to those used for time a shift in the origin of all or some of the spatial coordinates will not change the value of l.

1.3 The homogeneity and the isotropy of space

Δt and l are examples of invariants. Formally, Δt is invariant with respect to translations of the origin of time, whilst l is invariant with respect to translations of the spatial origin. This arbitrariness in the origin of both space and time is sometimes referred to as the homogeneity of space and time respectively.

That the length, l, is the same when measured in a coordinate system which is rotated with respect to the first is easily proved (see problem 1, Chapter 1). This leads to the conclusion that there is no preferred direction in space. Space is also said, therefore, to be isotropic. Actually we have really inverted the argument; the physical assumption is that space is both homogeneous and isotropic. Hence we can use Euclidean geometry to show that length is an invariant. Similarly it is the physical assumption of the homogeneity of time which leads to the time interval being an invariant. However, the correctness of Euclidean geometry comes more naturally to most people than the more abstract concepts of homogeneity and isotropy of space.

The above discussion has gone into some detail to emphasize

that reference frames, on the assumptions of the homogeneity of space and time and the isotropy of space, are arbitrary. It follows therefore that any physical law must not depend upon the particular reference frame we have chosen to select. To give an example, the equations of motion of two particles in one dimension acting through a force $F(x_1, x_2)$ are

$$m_1 \frac{d^2x_1}{dt^2} = F(x_1, x_2)$$
$$m_2 \frac{d^2x_2}{dt^2} = -F(x_1, x_2) \qquad [1.4]$$

where x_1 and x_2 are the coordinates of particles 1 and 2 in a particular reference frame and m_1 and m_2 are their masses. Now suppose we describe the same two particles by using a reference frame whose origin is at x_0 in the original frame. Then if we call the coordinates of the particles in the new frame x_1' and x_2', then

$$x_1 = x_1' + x_0 \text{ and } x_2 = x_2' + x_0$$

Since x_0 is a constant, its derivative with respect to time is zero and hence if we substitute these transformation equations into the equations of motion (equations [1.4]) we get

$$m_1 \frac{d^2x_1'}{dt^2} = F(x_1' + x_0, x_2' + x_0)$$
$$m_2 \frac{d^2x_2'}{dt^2} = -F(x_1' + x_0, x_2' + x_0)$$

the equations of motion in the new reference frame.

Now there are two things wrong with this result.

1. The new equations depend upon the origin of the coordinate system, x_0, which clearly offends against the homogeneity of space.
2. The force of interaction does not have the form $F(x_1', x_2')$, which it should if the equations are to have the same form as in the original reference frame.

This latter point means that these equations of motion do not satisfy the principle of relativity.

Now, earlier on we discussed the fact that, in going from one frame to another with a different origin, the invariant was not the absolute position of one event but the distance between two events. This suggests that if the forces between two particles always depended only upon the distance between the particles then we would remove the difficulties discussed above. In fact we find in nature that all forces between particles have this property. (Strictly speaking, although the magnitude of forces in nature depends upon their distance apart, their direction is not always along the line of centres. These are known as non-central forces.) Thus we can write

$$F(x_1, x_2) = f(x_1 - x_2)$$

and hence in the new reference frame the equations of motion become

$$m_1 \frac{\mathrm{d}^2 x_1'}{\mathrm{d}t^2} = f(x_1' - x_2')$$
$$m_2 \frac{\mathrm{d}^2 x_2'}{\mathrm{d}t^2} = -f(x_1' - x_2') \qquad [1.5]$$

These equations are now independent of x_0 and have the same form as the original equations.

The requirement that the equations of motion look the same in all reference frames is known as requiring that they be form invariant. Ensuring that this was so in the above example was in fact applying the principle of relativity. However, before explicitly giving and explaining this principle, there is one other transformation between coordinate systems that we should investigate.

1.4 Reference frames with a constant relative velocity

We have already agreed that space is homogeneous (as well as isotropic), i.e. if we displace the origin of the reference frame it should make no difference to any physical measurement or law. What happens, however, if we have a frame that is moving with respect to an original frame? Should length be an invariant and should the laws of physics be form invariant?

Before examining these questions it will be helpful to derive the transformation from the original reference frame to the moving frame. For so-called 'special relativity' we need only concern ourselves with the case when the relative velocity of the frames is constant. In particular, let the respective y and z axes of the two frames be parallel (because of the assumed isotropy of space we can always rotate one of the sets of axes to ensure that this is so). Further, let the x' axis of the second frame be moving along the x axis of the first frame with a constant velocity, $\boldsymbol{v}$. We shall refer to the original frame as the K frame and the moving frame as the K' frame. Alternatively, and somewhat more loosely, we shall refer to the laboratory frame and the moving frame. The corresponding situation in two dimensions is illustrated in Fig. 1.2.

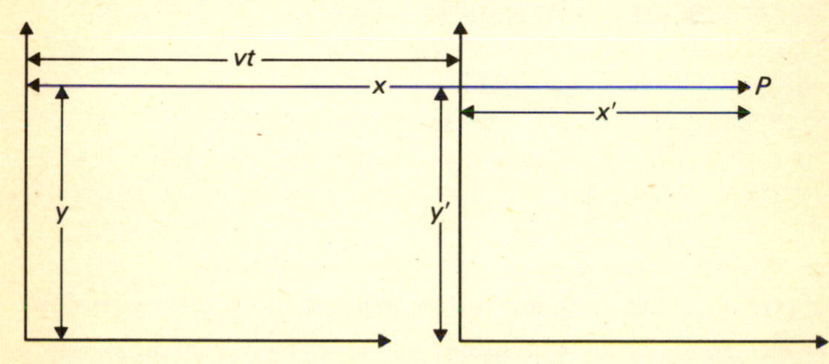

Fig. 1.2 The K' frame is travelling with a velocity along the x axis of the K frame. The y and y' axes are parallel. It is assumed that at $t = 0$, the origins are coincident and subsequently that time is the same in both frames, i.e. $t' = t$ for all t. The point P has the coordinates x, y in the K frame and x', y' in the K' frame. Geometry clearly shows that $y' = y$ and $x' = x - vt$.

The origins are coincident at $t = 0$ and we make the further assumption that time is the same in K as K'. The equations of the transformation, known as the Galilean transformation, are, from Fig. 1.2,

$$x' = x - vt \qquad y' = y \qquad z' = z \qquad [1.6]$$

The last of these equations comes from the analogy with $y' = y$, i.e. it is an axis at right angles to the direction of motion. It follows immediately, by differentiating with respect to time, that the equations for transforming the velocities in the K frame to their corresponding velocities in the K' frame are

$$u_x' = u_x - v \qquad u_y' = u_y \qquad u_z' = u_z \tag{1.7}$$

A second differentiation shows that the acceleration in the two frames is equal (remember that v is a constant), i.e.

$$a_x' = a_x \qquad a_y' = a_y \qquad a_z' = a_z \tag{1.8}$$

Now the question is: Can we distinguish between these two frames? We have already assumed that space is homogeneous and, given this, it would seem that the frames K and K' are indistinguishable. All we can say is that they have a relative velocity v. This is equivalent to making a small displacement $\mathrm{d}x$ in a time $\mathrm{d}t$. The K' frame moving with a constant velocity v corresponds to a series of displacements $\mathrm{d}x$ each in time $\mathrm{d}t$, such that $\mathrm{d}x = v\mathrm{d}t$. From this point of view the equivalence of K and K' is expressing nothing more than the homogeneity of space. However, this is not the whole story.

Suppose that in some sense space is absolute, i.e. it exists in its own right, or, to put the proposition another way, an object occupies a particular portion of space and if it is moved it no longer occupies that piece of space but another and different portion. Newton believed in such an absolute space and devised several ingenious experiments to demonstrate its existence. Other philosophers such as Leibniz and Mach denied the existence of absolute space and preferred to think of space as a series of relationships between objects. For the moment let us take Newton's side in the debate. Then it is clear that frames moving with a relative velocity can be distinguished, for there exists a reference frame in which absolute space is at rest. All other frames can be distinguished by their velocity with respect to the absolute frame. So if we have two frames moving along the x axis of the absolute frame with velocities v_1 and $v_1 + v$, then their relative velocity is v. Nevertheless, they can be distinguished because one has a velocity v_1 and the other $v_1 + v$ with respect to absolute space.

On the other hand, if we follow Leibniz and Mach and dispense with absolute space, then all we can say is that the two frames have a relative velocity, and whether one is moving and the other stationary or both are moving is a meaningless question. A description of the experiments which have been devised to settle the existence or non-existence of absolute space is not possible in this book. However, it is clear that Newton's equations, of which equation [1.5] is an example, do not contain the velocity and cannot therefore be used to identify absolute space. Its possible existence has been mentioned because it played such a dramatic role in the development of the theory of relativity. That role is described in Chapter 2.

For the moment we need not worry about the existence or non-existence of absolute space, but can go back to the Galilean transformation. Like the coordinate systems related by a linear displacement, length is an invariant, thus

$$l'^2 = (x_2' - x_1')^2 + (y_2' - y_1')^2 + (z_2' - z_1')^2$$

Using the transformation equations [1.6] yields $l' = l$, where l is given by equation [1.3]. We have, of course, assumed that the time interval between two events is also an invariant.

To summarize, physicists assume that space is homogeneous, isotropic and, nowadays, that there is no absolute space.

1.5 Inertial frames

All this is very interesting but if it is to be of any use then physical, as well as possible philosophical, consequences must flow. What these consequences are will preoccupy us till the end of this book. Before we can go on and look at them, however, we have to do two things. First, we have to describe more fully what we mean by a reference frame that is suitable for describing physical phenomena. Second, we must formulate precisely, rather than the loose description that we have used up till now, the principle that follows from the properties of space listed above.

We have already specified what we mean by a reference frame, but the question remains: Will all reference frames be equally suitable for formulating the laws of physics? Consider a reference frame which has its origin at the centre of the Sun and the direction of its axes are fixed with respect to the fixed stars. (A hundred years ago the fixed stars would be those that could be observed with the naked eye. Today it is possible to use the distant galaxies to determine a frame for say the solar system.) This reference frame is the one which is usually chosen to describe the motion of the planets. In this frame the planets have elliptical orbits. Now consider a reference frame which has the same origin as the first but whose axes are rotating with the angular velocity of the Sun. In order to describe the motion of the planets we would need a much more complicated set of equations; the forces that we would need would not be simply the inverse square law of the normal gravitational forces. Similarly, if we had a coordinate system whose origin was the Earth but whose axes were always inclined to the fixed stars, the Earth's motion would be very simple, i.e. it would be at rest. The motion of the other planets and the Sun would now be very complicated (since the Greeks adopted this coordinate system they had great difficulties describing the motion of the rest of the solar system).

From these examples it would seem that there exists one reference frame, or class of reference frames, for which the description of the motion of the planets is much simpler than in other frames. The question is, then: How can we select this particular frame or set of frames?

In fact it turns out that a particle, or set of particles, only obeys Newton's laws in a particular class of reference frames. These frames are known as inertial frames and they are defined as follows:

> If, in a particular reference frame, an isolated particle moves in a straight line with a constant velocity, then that frame is said to be an inertial reference frame.

Suppose that we have identified one inertial frame and that we generate another frame by a simple translation in space. Since space is homogeneous, changing the origin will not change the velocity of the particle and hence we have generated a new inertial frame, since the speed of the particle is still constant and in the same direction. Similar considerations will apply to frames generated by rotations of the axes. Thus from one inertial frame, determined in principle by observing the motion of an isolated particle, we can generate an infinite set of inertial frames by using different combinations of displacements and rotations.

That, however, does not exhaust the possibilities for generating inertial frames. The Galilean transformation for velocities (equation [1.7]) shows that in the new frame an isolated particle would have a constant velocity if, in the original frame, it also had a constant velocity. The velocities in the two frames will be different but both will be constant. Since our definition of inertial frames only requires that the velocity should be constant and not have a particular value, the new frame is also an inertial frame. Thus there exists a further class of inertial frames which can be generated from one inertial frame by having frames which move with a constant velocity with respect to the first.

Having now defined the reference frames which we will primarily be interested in, we can make the central hypothesis of the subject of 'special relativity':

> It is impossible by using any physical law to distinguish between inertial frames.

We have already, by our assumptions of homogeneity, isotropy and the absence of an absolute space, implied the principle. What remains is to work out the implications. Both Galileo and Newton believed in the principle of relativity but they were restricted by only having to believe that it held for dynamical systems. Einstein's great contribution was to insist that it applied equally well to that other great triumph of nineteenth-century physics, electromagnetic theory. In the next chapter we examine the two pillars of pre-quantum physics and investigate the consequences of applying the principle of relativity.

Chapter 2
Mechanics and electromagnetism

2.1 Introduction

It was mentioned at the end of the last chapter that our intention now is to look at the theories of classical, or Newtonian, mechanics and electromagnetic theory. Although the former was established almost two hundred years before the latter, it did not enjoy the same rigorous foundations. One of the first to recognize this was Ernest Mach (1836-1916), who was to have such a great influence on Einstein. Since Mach's day, Newton's laws have been the subject of many philosophical investigations. We do not have space to detail these arguments, but on the other hand we must be clear precisely what the theory of classical mechanics is before we attempt to see whether it satisfies the principle of relativity or not. In the end, however, we will see that it is electromagnetic theory that will stand unaltered and that all our careful formulation of mechanics has to be dramatically modified. Fortunately in this modification there will be some concepts that, because we have taken the trouble to be precise, we will be able to take over to the correct formulation. Further it should be said that there are still large areas of physics where Newton's equations are a more than adequate approximation.

2.2 Newtonian mechanics

Newton based his analysis of mechanical systems on his three famous laws:

1. If there are no forces acting on a particle it will persist in its motion, that is it will move along a straight line with a constant velocity.

2. If there are forces acting on a particle, the rate of change of linear momentum is equal to the force acting on it.

3. Action and reaction are equal and opposite.

For a student meeting mechanics for the first time, the simplicity of these laws means that, given a force – i.e. given how the force depends upon the positions of the particles – he/she can immediately write down a set of second-order differential equations for the motion of the particles. Equations [1.5] are an example of such a set of equations, provided that f is a known function. In three dimensions we can use vector notation. For example, suppose the magnitude of the force between two particles, both of mass m and whose positions are $\mathbf{r}_1$ and $\mathbf{r}_2$, is $m\omega^2|\mathbf{r}_1 - \mathbf{r}_2|$, where ω^2 is a constant. Then the second and third laws give us the differential equations

$$m\frac{d^2\mathbf{r}_1}{dt^2} = -m\omega^2|\mathbf{r}_1 - \mathbf{r}_2|$$
$$m\frac{d^2\mathbf{r}_2}{dt^2} = m\omega^2|\mathbf{r}_1 - \mathbf{r}_2| \qquad [2.1]$$

The solutions to these equations can be written

$$(\mathbf{r}_1 + \mathbf{r}_2)/2 = \mathbf{V}t + \mathbf{R}$$
$$(\mathbf{r}_1 - \mathbf{r}_2)/2 = \mathbf{A}\cos(\omega t + \varphi) \qquad [2.2]$$

where $\mathbf{R}$, $\mathbf{V}$ and $\mathbf{A}$ are constant vectors and φ is a constant scalar. The physical interpretation of these results is obvious. The first equation tells us that the centre of mass is moving in a straight line with a constant velocity. The second tells us that the distance between the particles is oscillating with a frequency ω.

For most applications there is no need to examine the internal consistency of these laws. The fact is that an enormous amount of physics has been based upon them and the agreement with experiment in nearly all cases has been extremely good. To most physicists, if a law works, in the sense that it predicts or explains physical phenomena, then it is best left alone. Only if its predictions or explanations seriously conflict with experiment does it merit re-examination. Such a re-examination of classical mechanics took place at the turn of this century and led, on the one hand, to the birth of quantum mechanics and, on the other hand, to a generalized form of Newtonian mechanics. In the first case the re-examination became necessary because Newtonian mechanics turned out to be a very bad description of the microscopic world of the atom and its constituents. The reasons for the second development will come later in this book.

At the moment we are not concerned with revising Newton's laws but reformulating them. The reason that we have to reformulate them is that, as presented above, they mix laws and definitions. From a practical point of view, as we have argued above, this may not matter, but when we come to apply such a sweeping principle as the principle of relativity we have to be rather more precise as to what are definitions and what are laws. Mach attempted to formulate Newton's laws so that the difference between laws, which are based on experiment, and definitions, which are useful but arbitrary, is made. Many authors, acknowledged or not, have followed Mach's formulation, or some modification of it, and it is essentially his formulation that is presented here. (Mach's book *The Science of Mechanics* is now regarded as historically inaccurate, but nevertheless his clear analysis of the problem of mechanics still stands.)

Reference back to Chapter 1 and the definition of inertial frames shows that Newton's first law is really equivalent to this definition, i.e. if an isolated particle moves in a straight line then we have an inertial frame and we can apply Newton's laws. Once we have established one inertial frame, all other frames related to the first by a Galilean transformation will also be inertial, as discussed in Chapter 1.

The problem with Newton's second law is that it is also really only a definition, i.e. force is defined as rate of change of momentum or, what is the same thing, mass times acceleration.

The only real physics is in the third law, and for this reason we will reformulate the laws with this law, more precisely formulated and taking pride of place. As the third law stands, it says that action and reaction are equal and opposite but the problem is what do we mean by 'action' and 'reaction'? The usual interpretation is that the forces acting on the two particles are equal and opposite. At the moment, in our new formulation, force has not been defined. The obvious replacement is acceleration, particularly as the measurement of acceleration depends only upon the measurement of length and time, the two fundamental concepts that we agreed in Chapter 1 caused little or no confusion in their definition. We should also remember that we have not, as yet, defined 'mass', so we cannot replace force directly by mass times acceleration. Our reformulation is therefore as follows:

> If we have two isolated but interacting particles, then at any instant of time the ratio of the magnitude of their accelerations (when measured in an inertial frame) is a constant. Further, the accelerations are oppositely directed. Thus, if $\mathbf{a}_1$ and $\mathbf{a}_2$ are the accelerations of the particles, then
>
> $$a_1/a_2 = k_{12} \text{ (a constant)} \qquad [2.3]$$
>
> and $$\mathbf{a}_1/a_1 = -\mathbf{a}_2/a_2. \qquad [2.4]$$

If k_{12} were taken to be the ratio of the Newtonian masses m_1/m_2, and force were to be defined as mass times acceleration, then equations [2.3] and [2.4] would be identical to Newton's third law (equation [2.3] showing that the forces are equal in magnitude and equation [2.4] that they are oppositely directed). However, in order to define mass we need to take into account one further piece of experimental evidence:

> If each of the two particles is replaced in turn by a third, and if $a_3/a_1 = k_{31}$, then
>
> $$a_3/a_2 = k_{32} = k_{31}k_{12}. \qquad [2.5]$$

From this law a number of deductions can be made. Firstly it enables us to define mass. From equation [2.5] we have $k_{31}k_{12} = k_{32}$ and since k_{32} must be independent of the properties of particle 1, it follows by inspection that k_{12} must be a ratio of two constants, one of which depends upon particle 1 and the other on particle 2. These constants are conventionally known as the inertial masses. If the ratio is written

$$k_{12} = m_2/m_1 \quad [2.6]$$

then m_1 is the inertial mass of particle 1 and m_2 the inertial mass of particle 2. Similarly $k_{31} = m_1/m_3$, $k_{32} = m_2/m_3$, etc. Units of mass are obtained, at least in principle, by assigning unit mass to an arbitrary particle and the masses of other particles are then determined by the use of equations [2.3] and [2.6].

A combination of equations [2.3], [2.4] and [2.6] leads to the result

$$m_1\mathbf{a}_1 + m_2\mathbf{a}_2 = 0$$

A single integration over time yields the well-known law of the conservation of linear momentum for two particles

$$\mathbf{p}_1 + \mathbf{p}_2 = \mathbf{C} \text{ (a constant vector)} \quad [2.7]$$

where

$$\mathbf{p}_1 = m_1\mathbf{u}_1, \qquad \mathbf{p}_2 = m_2\mathbf{u}_2 \quad [2.8]$$

are defined as the momenta of the two particles and $\mathbf{u}_1$ and $\mathbf{u}_2$ and their respective velocities.

If we define the centre of mass for the two particles by

$$\mathbf{R} = (m_1\mathbf{r}_1 + m_2\mathbf{r}_2)/(m_1 + m_2) \quad [2.9]$$

where $\mathbf{r}_1$ and $\mathbf{r}_2$ are the positions of the particles, then a time integration of equation [2.7] gives the trajectory of the centre of mass, i.e.

$$\mathbf{R} = \mathbf{V}t + \mathbf{R}_0$$

where $\mathbf{V} = \mathbf{C}/(m_1 + m_2)$ and $\mathbf{R}_0$ are constant vectors. These results can be extended, with no further assumptions, to more than two particles.

Thus our reformulated law as given by equations [2.3] to [2.5] enables us to define mass, to derive the law of conservation of momentum and hence show that the centre of mass moves in a straight line. This latter result was an important one for Newtonian mechanics because it meant that the motion of the whole system of particles could easily be dealt with, and the only remaining problem was to determine their relative motion. Equations [2.2] are a simple example of this result.

We must now examine whether this law satisfies the principle of relativity. That it does is very easy to see; equation [1.8] showed that each component of the acceleration was unchanged under a Galilean transformation. Hence if we denote variables in the new frame K' by adding primes to the corresponding quantities in the frame K (the reader is urged to go back to Fig. 1.2 if he is still not clear as to what is meant by the frames K and K') we get

$$a_1'/a_2' = a_1/a_2 = k_{12}$$

where we have used equation [2.3]. Thus the ratio of accelerations in the new frame is also a constant (in fact the same constant) and thus this part of the law satisfies the principle of relativity. It is easily seen that similar considerations will establish that equations [2.4] and [2.5] also satisfy the principle. Since k_{12} has the same value in all frames, it follows that the mass of the particle will be the same in all inertial frames. Once we have established that the basic law satisfies the principle of relativity, it follows that the law of conservation of momentum and the straight-line trajectory for the centre of mass also hold in all inertial frames. The reader can check this by applying the Galilean transformation to these results (see exercise 3, chapter 1, Appendix 3).

In order to proceed further we need to define force: this is really Newton's second law. The definition is that the product of the mass of the particle and its acceleration is the force, i.e.

$$\mathbf{F} \equiv m\mathbf{a} \qquad [2.10]$$

The last law of motion can now be stated:

> There exists a law of force between two particles which at the most can only depend upon the difference between the positions of the particles and the difference of their velocities, i.e.
>
> $$\mathbf{F}_{ij} = \mathbf{F}_{ij}(\mathbf{r}_i - \mathbf{r}_j, \mathbf{u}_i - \mathbf{u}_j) \qquad [2.11]$$
>
> where $\mathbf{F}_{ij}$ is the force on particle i due to the presence of particle j. Further the total force on particle i due to the presence of all other particles is the sum of the individual contributions, i.e.

$$\mathbf{F}_i = \sum_j \mathbf{F}_{ij} \qquad [2.12]$$

where $\mathbf{F}_i$ is the total force on particle i.

Since $\mathbf{F}_{ij}$ depends only upon the differences of position and velocity, it follows from the equations of the Galilean transformation (equations [1.6] and [1.7]) that it has the same form in all inertial frames. Clearly we cannot use this law to distinguish inertial frames.

This finishes our reformulation of the laws of mechanics. Their authenticity, at least within a certain well-defined domain, is well established. Our next step is to look at the other great colossus of nineteenth-century physics: electromagnetic theory.

2.3 Electromagnetic theory

The situation in the middle of the nineteenth century was, therefore, that the mechanics of Newton were seen to be compatible with the principle of relativity. In that century James Clerk Maxwell (1831-79), building on the pioneering work of Faraday (1791-1867), Ampère (1775-1836) and others, formulated his famous laws of electromagnetism and it appeared that electromagnetism could join classical mechanics as the other basic foundation of physics. Further, it gave to the science of optics a sound theoretical base, showing that light was nothing more than an electromagnetic wave. It should now be obvious to the reader that the next step was to test these equations against the principle of relativity, that is to say, if the Galilean transformation is applied to Maxwell's equations, do they remain form invariant, i.e. is the form of the equations preserved under the transformation?

By the same token that we argued that it was necessary to formulate Newton's laws rather carefully before applying the principle of relativity, we must also look carefully at Maxwell's equations. We start by defining the electric and magnetic fields. Suppose we have a charged particle which is instantaneously at rest. Then if that particle experiences a force, i.e. it accelerates, it is said to be under the influence of an electric field. This field is a vector quantity whose direction is in the direction of the force and whose magnitude at a point is defined as the force acting on a unit charge at that point. Thus if we denote by $\mathbf{\mathcal{E}}$ the electric field acting on a particle with a charge q, and denote the force acting on the particle by **F**, then

$$\mathbf{\mathcal{E}} = \mathbf{F}/q \quad [2.13]$$

(We could have a problem about defining charge since the only obvious manifestation of charge is its property of allowing the particle to be influenced by an electric field. However, we can define the charge in terms of the force between two charges rather than the force on a single charge due to an electric field.)

Two things need to be said about the definition in equation [2.13]. First, it must be remembered that force is a shorthand way of saying mass times acceleration. Thus to measure the electric field at any point we have to measure the acceleration of a charged particle. Second, the field can exist anywhere in space and, furthermore, its magnitude and direction can vary both in space and time.

Having now defined the electric field, we have to have a law which governs its behaviour in both space and time. It was known long before the time of Maxwell that the sources of electrostatic fields were themselves charges. The inverse square law governing the force on one charge due to another was given by Coulomb (1736-1806), whose name it now bears. Subsequently Gauss (1777-1855) showed that it followed from Coulomb's law, that the total flux coming from a volume was equal to the total charge enclosed. In turn, from Gauss's integral law, it is possible to derive a differential form of the law:

$$\nabla \cdot \mathbf{\mathcal{E}} = \rho/\epsilon_0 \quad [2.14]$$

where ρ is the charge density and ϵ_0 is the permittivity of free space. Equation [2.14] is the first of Maxwell's equations. Thus, just as the dynamical law expressed in equations [2.3] and [2.4] led to the conservation of momentum and subsequently to the linear trajectory of the centre of mass, a hierarchy of different statements of the same law flows from the empirical formulation of Coulomb.

The definition of the magnetic field is slightly more difficult to formulate. For simplicity let us assume that when the charged particle is at rest there is no force acting on it, i.e. there is no electric field present. If the particle is now given a constant velocity **u** it may be that the particle experiences a force. If it does then there is said to be a magnetic field present. The force which acts on the particle is an unusual one compared to the forces we have met so far. First, it acts in a plane that is at right angles to the direction of **u**. Second, the magnitude of the force depends linearly on the magnitude of **u** and, furthermore, this magnitude varies sinusoidally as the direction of **u** is rotated in the plane perpendicular to the direction of the force. The situation is illustrated in Fig. 2.1. We define the magnetic field **B**, which, like the electric field, is a vector quantity, as follows:

$$\mathbf{F} = q\mathbf{u} \times \mathbf{B} \qquad [2.15]$$

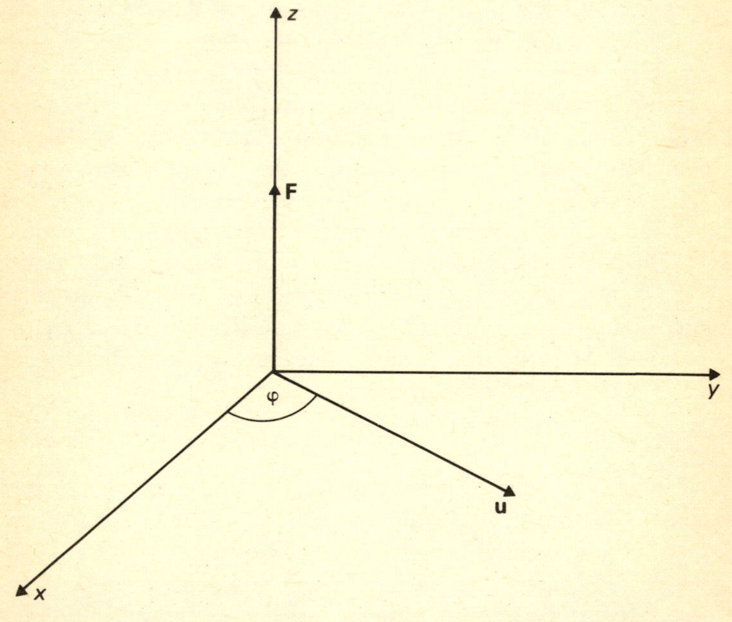

Fig. 2.1 The velocity vector **u** of the particle lies in the *x*, *y* plane. The direction of the force due to the magnetic field is in the *z* direction. The magnitude of the force varies sinusoidally with the angle.

Since in a vector product the magnitude is proportional to the sine of the angle between the two vectors and its direction is perpendicular to both of them, this definition leads to the force **F** having the properties listed above. Experiment shows that the magnetic field obeys the equivalent of Gauss's theorem for the electric field. However, there is one important difference. Despite attempts to search for magnetic monopoles, to date none has been found, although they are predicted by some theories of elementary particles. Thus although there is a magnetic analogue of equation (2.14), the magnetic monopole density (which is the analogue of the charge density) is always zero. The differential equivalent of Gauss's law is therefore

$$\nabla \cdot \mathbf{B} = 0 \qquad [2.16]$$

If we combine the definitions of the electric and magnetic fields, then the total force acting on a particle with a charge q and moving with a velocity **u**, is

$$\mathbf{F} = q\,(\boldsymbol{\mathcal{E}} + \mathbf{u} \times \mathbf{B}) \qquad [2.17]$$

This force is known as the Lorentz (1853-1928) force. It should be emphasized that equation [2.17] is a physical law, even though it has been used to define the fields. This is because it expresses the empirical facts that: the force is proportional to the charge, has a constant term and a term proportional to the velocity. We have used the law to define the fields, so we must now examine whether these definitions lead to results at odds with the principle of relativity.

The first question to ask is do the fields, as defined in this way, have the same value in all inertial frames? The answer is clearly no. Consider a particle which has an instantaneous velocity u in the x direction (since the force is defined as mass times acceleration, the velocity of the particle must be continually changing) and suppose that, using the definitions given above, there is an electric field $\mathcal{E}_y$ in the y direction and a magnetic field B_z in the z direction. Then according to equation [2.17] above the total force acting on the particle will be in the y direction and its magnitude will be

$$F = q(\mathcal{E}_y - uB_z) \qquad [2.18]$$

Now consider a frame which is moving along the x axis of the first frame with a velocity u, i.e. as shown in Fig. 1.2 (except that the relative velocity of separation is u rather than v). As before, we shall refer to the first frame as the K frame and the second as the K' frame. In the K' frame the particle is at rest and, by definition, the only force acting upon it can be that due to the electric field. Now we have seen above that both mass and acceleration have the same value in all inertial frames. Thus since the force on the particle is defined as their product, its value in the K and K' frames must be the same. It follows, therefore, that the electric field in the K' frame must be given by

$$\mathcal{E}'_y = \mathcal{E}_y - uB_z \qquad [2.19]$$

which leads to a magnitude of $\mathcal{E}'_y$ that is manifestly different from $\mathcal{E}_y$. Equation [2.19] follows from the fact that in K

$$ma = q(\mathcal{E}_y - uB_z)$$

whereas in K'

$$ma' = q\mathcal{E}'_y$$

and according to equation [1.8] $a' = a$. It would appear, therefore that we could distinguish between inertial frames by saying that in one frame, K' there is both an electric and a magnetic field, whilst in the other frame, K', there exists only an electric field. This would, however, be wrong because one must remember that the only observable is the acceleration. Convenient though it is to define the electromagnetic field, it is a hypothetical construct, derived from the measurement of its effect.

Having satisfied ourselves that the definitions do not conflict with the principle of relativity, we can make use of equation [2.19] to relate the y component of the electric field in the K' frame to the appropriate fields in the K frame. It can be seen that what is an electric field in one frame is an admixture of electric and magnetic fields in another. Thus the distinction between the two is somewhat artificial. Nevertheless a compromise can be reached by referring to the two as the electromagnetic field. This result is not often presented, and so the mixing of electric and magnetic fields in different inertial frames is sometimes taken as if it were a result of post-Einsteinian theory. As we have seen this is not the case; in a Galilean transformation the fields are mixed as we transform from one inertial frame to another, as they are in Einstein's theory. In the latter case the admixture is more complicated than in the Galilean case, only the x components of the electric and magnetic fields being unchanged in the transformation (see Chapter 7). Exercise 5 (Chapter 2, Appendix 3) asks the reader to derive the relationships between the other components of the field. The actual relationships are

$$\begin{aligned} B'_x &= B_x, \quad B'_y = B_y, \quad B'_z = B_z, \\ \mathcal{E}'_x &= \mathcal{E}_x, \quad \mathcal{E}'_y = \mathcal{E}_y - uB_z, \quad \mathcal{E}'_z = \mathcal{E}_z + uB_y \end{aligned} \qquad [2.20]$$

To find the inverse transformation, we could of course solve these equations to find $\mathcal{E}_x, B_x$, etc., as functions of the fields in the K' frame. However, it is convenient to make use of a trick that we will be able to use for other quantities related by the Galilean transformation. Suppose we regard the K' frame as the laboratory frame. Then the K frame becomes the moving frame and equations [2.20] apply, except that we must now interchange primed and unprimed quantities and change u to –u (moving frame has a velocity –u with respect to the laboratory frame). Thus the first four of equations [2.20] remain the same but the fifth and sixth become

$$\mathcal{E}_y = \mathcal{E}'_y + uB'_z \qquad \mathcal{E}_z = \mathcal{E}'_z - uB'_y \qquad [2.21]$$

Finally we come to the last two of Maxwell's equations. The first of them is a modification of Ampère's law, which states that the line integral of the magnetic field round a closed path is proportional to the total current flowing through the enclosed area. Maxwell argued that there was an alternative source of current which could be attributed to the presence of a time-varying electric field. Known as the 'displacement current', it was simply added to the current term to give a modified form of Ampère's integral law. Like the first two of Maxwell's equations, there is an alternative formulation in the form of a partial differential equation:

$$\nabla \times \mathbf{B} = \epsilon_0 \mu_0 \frac{\partial \boldsymbol{\mathcal{E}}}{\partial t} + \mu \mathbf{J} \qquad [2.22]$$

where μ_0 is the permeability of free space and **J** is the current density. Finally we have Faraday's induction law, which has, like the others, both integral and differential forms. The latter is:

$$\nabla \times \mathbf{\mathcal{E}} = -\frac{\partial \mathbf{B}}{\partial t} \qquad [2.23]$$

2.4 Maxwell's equations and the principle of relativity

Before applying the principle of relativity it is convenient to make a simplification and look just at the properties of Maxwell's equations in free space, i.e. where there are no source terms present. Thus in equation [2.14] we can put $\rho = 0$ and in [2.22] $\mathbf{J} = 0$. In free space equations [2.14], [2.16], [2.22] and [2.23] consequently can be written

$$\begin{aligned} \nabla \cdot \mathbf{\mathcal{E}} &= 0 & \nabla \cdot \mathbf{B} &= 0 \\ \nabla \times \mathbf{B} &= \epsilon_0 \mu_0 \frac{\partial \mathbf{\mathcal{E}}}{\partial t} & \nabla \times \mathbf{\mathcal{E}} &= -\frac{\partial \mathbf{B}}{\partial t} \end{aligned} \qquad [2.24]$$

In order to transform these equations to the frame K', we have to transform the operators $\partial/\partial t$, $\nabla \cdot$ and $\nabla \times$ from the coordinates x, y, z to the coordinates x', y', z'. Consider first the operator $\partial/\partial t$. Let this operate on some function of x, y, z, t, say $f(x, y, z, t)$. Now x', y', z', t' are related to the original coordinates by the inverse of the Galilean transformation (equation [1.6]). The inverse transformation can be obtained by arguments similar to those used above for the inverse of the transformation of the electromagnetic field. Thus

$$x = x' + vt' \quad y = y' \quad z = z' \quad t = t' \qquad [2.25]$$

Thus we can write

$$f = f(x' + vt', y', z', t') = g(x', y', z', t')$$

and hence

$$\frac{\partial f}{\partial t} = \frac{\partial t'}{\partial t}\frac{\partial g}{\partial t'} + \frac{\partial x'}{\partial t}\frac{\partial g}{\partial x'} + \frac{\partial y'}{\partial t}\frac{\partial g}{\partial y'} + \frac{\partial z'}{\partial t}\frac{\partial g}{\partial z'}$$

$$= \frac{\partial g}{\partial t'} - v\frac{\partial g}{\partial x'}$$

where we have used the Galilean transformation to obtain the partial derivatives $\partial t'/\partial t$, $\partial x'/\partial t$, etc. The equivalent result for the operator is

$$\frac{\partial}{\partial t} = \frac{\partial}{\partial t'} - v\frac{\partial}{\partial x'} \qquad [2.26]$$

The other partial derivatives can be transformed in a similar manner (see exercise 6). The results are

$$\frac{\partial}{\partial x} = \frac{\partial}{\partial x'} \qquad \frac{\partial}{\partial y} = \frac{\partial}{\partial y'} \qquad \frac{\partial}{\partial z} = \frac{\partial}{\partial z'} \qquad [2.27]$$

We are now formally in a position to look at the transformation properties of Maxwell's equations. Rather than involve ourselves in a large amount of cumbersome mathematics in treating the general case, however, we will examine the special case where the electric field is always in the y direction and the magnetic field is in the z direction. Further we shall assume that the spatial dependence of both of these fields is in the x direction only; they both are assumed to be time dependent. A particular example of such a field configuration would be a plane-polarized wave travelling in the x direction. Under these assumptions the first two of Maxwell's equations in free space (equations [2.24]) reduce to identities and the second two become

$$-\frac{\partial B_z}{\partial x} = \epsilon_0\mu_0\frac{\partial \mathcal{E}_y}{\partial t}, \quad \frac{\partial \mathcal{E}_y}{\partial x} = -\frac{\partial B_z}{\partial t} \qquad [2.28]$$

If we look for solutions of the form

$$\begin{aligned} \mathcal{E}_y &= \mathcal{E}_0 \exp\{i(\omega t - kx)\} \\ B_z &= B_0 \exp\{i(\omega t - kx)\} \end{aligned} \qquad [2.29]$$

substitution in the above equations gives

$$\begin{aligned} kB_0 &= \omega\mathcal{E}_0/c^2 \\ k\mathcal{E}_0 &= \omega B_0 \end{aligned}$$

where

$$c^2 = 1/(\epsilon_0\mu_0) \qquad [2.30]$$

The solution to these equations is simply obtained to give

$$\mathcal{E}_0/B_0 = \pm c \qquad ck = \pm \omega \qquad [2.31]$$

The two signs correspond to the two directions of propagation along the x axis. This, then, represents the plane-polarized wave mentioned earlier and gives Maxwell's celebrated result for the velocity of propagation for electromagnetic radiation.

To test the form invariance of Maxwell's equations we have to transform equations [2.28] using the transformation equations [2.26], [2.27]. Thus

$$-\frac{\partial B_z}{\partial x'} = \frac{1}{c^2}\frac{\partial \mathcal{E}_y}{\partial t'} - \frac{v}{c^2}\frac{\partial \mathcal{E}_y}{\partial x'}$$

$$\frac{\partial \mathcal{E}_y}{\partial x'} = -\frac{\partial B_z}{\partial t'} + v\frac{\partial B_z}{\partial x'}$$

These equations still involve the fields as measured in the K frame, so finally we need to use equations [2.20] and [2.21] which relate these fields to those measured in the K' frame. The result is

$$\frac{\partial \mathcal{E}'_y}{\partial x'} = -\frac{\partial B'_z}{\partial t'}$$

$$-(1 - v^2/c^2)\frac{\partial B'_z}{\partial x'} + \frac{v}{c^2}\frac{\partial \mathcal{E}'_y}{\partial x'} = \frac{1}{c^2}\frac{\partial \mathcal{E}'_y}{\partial t'} + \frac{v}{c^2}\frac{\partial B'_z}{\partial t'} \qquad [2.32]$$

The first of these equations clearly has the same form as its corresponding equation [2.28] in the K frame, but what is equally clear is that the second one is not form invariant. This is a problem; we must apparently either abandon electromagnetic theory or the principle of relativity! Neither proposition is particularly attractive.

Maxwell's equations were the culmination of long years of scholarship and seemed to fit with all the known facts, as well as predicting new and exciting phenomena such as the description of light as an electromagnetic wave. On the other hand we have consistently argued that the principle of relativity is fundamental, implying both the homogeneity of space and the absence of an absolute reference frame. The first of these implications seems to be fundamental. The second could be abandoned; Newton wanted to have an absolute space, but, because they involved the acceleration, his equations were unable to distinguish between relative and absolute space. If there were an absolute space, it would be compatible with Newton's laws and the question then remains: Would it help to solve the problem that Maxwell's equations do not appear to be form invariant under the Galilean transformation?

To try and get some insight into the problem we can try looking for wave-like solutions, such as those given by equation [2.29]. Substitution in equation [2.32] yields the equations

$$(1 - v^2/c^2)kB_0 - vk\mathcal{E}_0/c^2 = \mathcal{E}_0\omega/c^2 + v\omega B_0/c^2$$
$$\mathcal{E}_0/B_0 = \omega/k$$

After a little algebra these yield the solution

$$\begin{aligned} \mathcal{E}_0/B_0 &= (\pm c - v) \\ \pm\omega &= (c \pm v)k \end{aligned} \qquad [2.33]$$

If $v = 0$, then these solutions reduce to the equations [2.31] in the frame K.

The physical meaning of equations [2.33] is clear; the wave in the frame K' has a frequency which has been Doppler-shifted by an amount vk. The Doppler effect is most commonly observed in sound waves where the waves have to propagate through a medium, frequently air. If an observer is moving with respect to the medium, then he/she will detect a changed frequency compared to the frequency observed when he/she is stationary with respect to the medium.

This, then, gives us a possible way out of the dilemma of the lack of form invariance of Maxwell's equations. If we return to Newton's idea of an absolute space, which at the end of the last century became known as the 'aether', then this provides a 'medium' of propagation for the electromagnetic waves. In another reference frame which is moving with respect to the aether, we would expect to see a Doppler-shifted frequency of precisely the amount predicted by equations [2.33]. Further, we have a means of checking experimentally the validity of this statement. If we could find two different inertial frames that were moving with a relative velocity, then our proposition of an absolute space, or aether, should lead to the velocity of light being different in the two frames, since the phase velocity of the waves is ω/k (see equations [2.33]). If this were in fact so, then everything would be consistent and we would not have to abandon the homogeneity of space, which Newton's equations depend on, but simply accept the existence of absolute space. The experimentalists of the late nineteenth and early twentieth centuries then set themselves the task of detecting the aether by the measurement of the velocity of light in two different inertial frames.

The most celebrated of these experimentalists was an American physicist, A. A. Michelson (1852-1931), who for the most famous of his experiments was joined by another American, E. W. Morley (1838-1923). They used the Earth as an inertial frame, which, as we discussed in Chapter 1, can be used in this capacity provided we restrict the time of measurement to be small compared with the ratio of the velocity of the Earth to its acceleration. The velocity of the Earth is about 30 km s^{-1} and its period of rotation about the Sun is 1 year. Its angular velocity is therefore 2π radian $year^{-1}$. The linear acceleration is the product of the linear and angular velocities and hence the ratio of the linear velocity to the acceleration is 1/(angular velocity) = $1/2\pi \doteq 1/6$ year. Thus, provided the measurements take a time which is much less than 2 months, the Earth is an accurate inertial frame.

What Michelson and Morley did was to measure the velocity of light both parallel and perpendicular to the Earth's motion. They repeated this experiment at different times of the year; thus on one occasion the velocity of light perpendicular to the motion would be compared with its value parallel to the Earth's motion and in another case with its value anti-parallel to its motion.

In none of their experiments were they able, within the accuracy of their experiment, to detect any difference in the values of the velocity. Their method was an interferometric one, not giving the absolute values of the velocities but only their relative difference. Michelson and Morley used an interferometer with arms of equal length and it was soon pointed out that, if the Lorentz-Fitzgerald contraction was valid (see Chapter 1), this would contract the arm in the direction of the Earth's motion and one would not expect to see any shift in the interference fringes, since any shift due to different velocities in the two different directions would be exactly compensated by the different arm lengths. There is some validity in this criticism and nowadays the Michelson-Morley experiment is interpreted as demonstrating the isotropy of space. It was not until some 35 years later, when Kennedy and Thorndike repeated the experiment under conditions of greater stability and with an interferometer whose arms were of unequal length, that it was shown, again within the limits of experimental error, that the velocity of light was the same in different inertial frames. (For a detailed description of these experiments see Taylor and Wheeler, and Whittaker. For a formal discussion of the interpretation of them see H. P. Robertson, *Reviews of Modern Physics*, vol. 21, p. 378, 1949.)

The dilemma is glaringly obvious and the assumption of absolute space, or aether, is not going to save us; all the experimental evidence indicates that the velocity of light has the same value in all inertial frames. There is no Doppler shift as predicted by equation [2.33]. We must now decide whether to abandon Maxwell's equations, with all the weight of experiment behind them, or to abandon the principle of relativity, supported as it is by the equally tried and tested Newton's laws. We should be perhaps a little more precise about the last question; we really cannot abandon the principle of relativity without flying in the face not only of experiment but also of common sense. We would have to believe that in some sense space is 'different' as we move from place to place. However we could be getting incorrect results if the transformation that we are using to apply the principle of relativity is incorrect, i.e., simple though its derivation appeared, possibly the Galilean transformation is incorrect.

The only guide that we have is that an examination of Maxwell's equations in the K' frame (equation [2.32]) shows that if the relative velocity of the two frames v is small compared to the velocity of light, then they reduce to the same form as those in the K frame. This suggests that if we are talking about velocities that are much smaller than the velocity of light, then the Galilean transformation may be an adequate approximation. Given that up to the turn of the century the only dynamics that had been tested experimentally were the dynamics of macroscopic particles, whose velocities were minute compared to c, it is a real possibility that the Galilean transformation is only an approximation, any differences from which could not be detected by any dynamical experiment. Nevertheless, we do have a dilemma – electromagnetism or the Galilean transformation – the resolution of which will occupy the next chapter.

Chapter 3
Space and time

3.1 Introduction

The dilemma mentioned at the end of the last chapter was one of a small number of outstanding problems that faced physicists at the end of the last century. Among the others were the so-called 'ultra-violet catastrophe' and the photoelectric effect. Within the next two decades all of these problems, if not solved, would at least have the foundations of the solutions laid. In two of these problems the man responsible for the major breakthrough was Albert Einstein (1879-1955). It is difficult, after a span of some seventy years, to realize the magnitude of the problem and to appreciate the enormous intellectual and imaginative leap that Einstein had to make. It is, however, salutory to read that even for Einstein the problem was not an easy one.

> I realised that this difficulty was really hard to solve. I spent almost a year in vain trying to modify the ideas of Lorentz in the hope of resolving the problem.

For Einstein the solution came in a flash of inspiration that distinguishes genius from lesser men.

> By chance a friend of mine (Michele Besso) helped me out. It was a beautiful day when I visited him with this problem. I started the conversation in the following way: 'Recently I have been working on a difficult problem. Today I come here to battle against that problem with you.' We discussed every aspect of this problem. Then suddenly I understood where the key to this problem lay. Next day I came back to him and

said, without even saying hallo, 'Thank you. I've completely solved the problem!' (Quoted in S.Y. Ono, *Physics Today,* **35**, no. 8, p. 45, 1982.)

We can, of course, never know the thought processes that went on in Einstein's mind, so we cannot know the way he arrived at the solution. In any case, if the average physicist relied solely on intuitive flashes to solve every problem he would solve very few. Part of his armoury is a number of techniques that help to clarify the problem and to reduce it to its essentials. One such technique is to look in a formal manner at any derivations that may be involved. Thus if we wish to examine the validity of the Galilean transformation, then we write down, first of all, the assumptions that we are going to make and then look at the deductions. In this way we may hope to detect any logical flaw in the derivation.

3.2 Formal derivation of the Galilean transformation

The Galilean transformation can be derived on the basis of two assumptions.

1. The principle of relativity.
2. That time is the same in all reference frames, i.e. if t is the time in the K frame and t' the time in the K' frame then, if the clocks are synchronized at $t = t' = 0$, $t = t'$ always.

In addition to these two assumptions, we also need the concept and definition of velocity. In Appendix 2 a derivation based solely on these two assumptions is given. To avoid detracting from the main line of the argument, we shall make the additional assumption in this chapter that the transformation is linear. This additional assumption takes away the necessity of doing some rather formal mathematics.

Consider first the implications of 1 above. Suppose we have two reference frames, related as shown in Fig. 1.2, i.e., moving relative to one another along their common x axis. The initial conditions have been specified and the clocks synchronized so that at $t = t' = 0$ the spatial origins are coincident. Now suppose at the origin of time, when the y and y' axes are coincident, we make a mark on the y' axis with a marker fixed to the y axis. We now measure in the K frame the y coordinate of the marker and in the K' frame the y' coordinate of the mark. Now the principle of relativity tells us that the two measurements must be identical since, if they were different, it would be possible to distinguish between inertial frames, i.e. we would be able to label the frames according to the magnitude of the y coordinate. We could clearly carry out this procedure for all inertial frames and this would be a clear contradiction of the principle of relativity. We are therefore led to conclude the result:

$$y = y', \qquad z = z' \tag{3.1}$$

where the last result follows by analogy with the first. That is all the principle of relativity can tell us.

Next let us invoke the simplifying but unnecessary assumption that the transformation is linear, i.e. let us write

$$\begin{aligned} x' &= ax + bt \\ t' &= ex + ft \end{aligned} \tag{3.2}$$

where a, b, e and f are constants. From these equations we can derive the equations for transforming a velocity from one frame to another:

$$u_x' = \frac{\mathrm{d}x'}{\mathrm{d}t'} = \frac{a\mathrm{d}x + b\mathrm{d}t}{e\mathrm{d}x + f\mathrm{d}t} = \frac{a\mathrm{d}x/\mathrm{d}t + b}{e\mathrm{d}x/\mathrm{d}t + f}$$

or $$u_x' = \frac{au_x + b}{eu_x + f} \tag{3.3}$$

where u_x is the x component of the velocity in the K frame and u_x' the corresponding quantity in the K' frame. Now for a particle in the K frame, which has a velocity v, we know by definition that its velocity in the K' frame is zero if $u_x = v$, i.e. if $u_x = v$ then $u_x' = 0$. Hence from equation [3.3]

$$av + b = 0 \tag{3.4}$$

Similarly if $u_x = 0$ then $u_x' = -v$, hence

$$-v = b/f \tag{3.5}$$

Substitution of equations [3.4] and [3.5] into the transformation [3.2] gives

$$\begin{aligned} x' &= a(x - vt) \\ t' &= ex + at \end{aligned} \qquad [3.6]$$

Finally we invoke assumption 2 above, i.e. $t' = t$, and the second of equations [3.6] then tells us that $e = 0$ and $a = 1$. The first of equations [3.6] then gives the Galilean transformation [1.6].

With this formal analysis it is quite clear where the error, if any, must lie. We cannot give up the principle of relativity, and the definition of velocity is just a definition. All that remains is the assumption 2, that time is the same in all inertial frames. This is something that we have assumed implicitly from the beginning and which has always seemed intuitively obvious (and which, as we shall see, goes to show that intuition can sometimes be very misleading). If we are to question this assumption then we must examine rather carefully what we mean by time and how we measure it.

3.3 The concept and measurement of time

It still seems reasonable that time is homogeneous in any one reference frame, so that the choice of origin cannot be important. Thus the assumption above that time is the same in all reference frames should more correctly be that the time interval between two events is the same in all reference frames. If this assumption is not correct then we somehow have to explain why the same two events, if observed from different frames, will occur with different time intervals between them.

We stated in Chapter 1 that the concept of time was one that was familiar to the earliest societies; the simple observation of the rising of the Sun and its setting in the evening enabled man to count the time in days. In trying to obtain subdivisions of the day, however, we have to be careful. Suppose that we insisted that between every sunrise and sunset there were exactly twelve hours. Then we would find that the human pulse would run slower in the winter than it did in the summer. With no other guidance, we might try to explain this by some sort of physiological effect due to, say, changes in the temperature. As our science became more sophisticated we might be able to measure the frequency of an emission line from, say, hydrogen gas. Again we would find that the frequency would vary over the year, the highest frequency being observed in the summer and the lowest in the winter. Now in principle there is no reason why we should not base a physics on this definition of time, although the periodic phenomena are something we would have to explain.

An alternative definition of the unit of time could be based on the rotation of the Earth about its own axis with respect to the fixed stars. On this definition the pulse rate of most humans would have roughly the same value in the winter as in the summer. Now, however, the days, as defined by the time between sunrise and sunset, would not be equal. We no longer have to explain something about the physiology of the human body but we do have to explain something about the rotation of the Earth with respect to the Sun, i.e. to explain the variation of the length of day. This definition of time was used by Newton and it was his theory that solved the problem of the variation of the length of the day.

Suppose we now return to the measurement of the frequency of the emission line of hydrogen. There is now no detectable difference in the line frequency between summer and winter. But if we make very accurate measurements we will find there is a very small increase in frequency as time goes on. We can attempt to set up a theory to explain this phenomenon or, alternatively, we can use the frequency to define a unit of time and then say that the Earth is slowing down. This is in fact the procedure adopted in modern standards practice. The actual line used is not in the hydrogen spectrum but one in the spectrum of caesium. This particular line is defined as having a frequency of 9,192,631,770 Hz, where the hertz is the number of cycles per second. Thus, formally, the time elapsed in counting that number of oscillations is defined as the second. (It is possible to buy a commercial caesium clock in which the time is displayed in digital form. For an accuracy of one part in 10^{12} the clock is about the size of a suitcase; if one is content with one part in 10^9, it is only the size of a book.)

Having defined the second in this way, how can we construct a

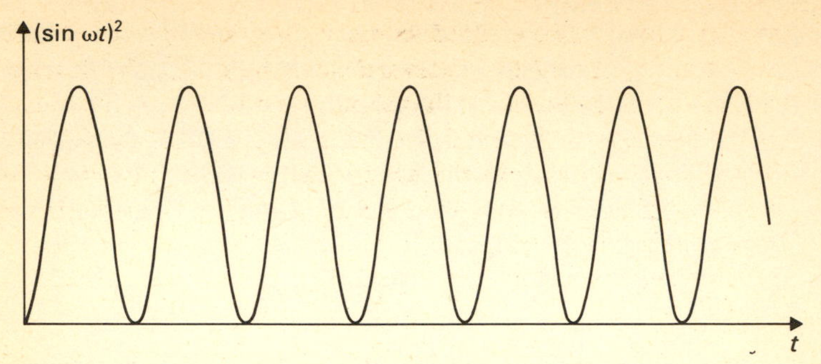

Fig. 3.1 The intensity of a pure sine wave as a function of time.

clock to measure time that will incorporate this definition? The intensity of a pure sine wave is shown in Fig. 3.1. Suppose that we construct a device that cuts out all the pulses except the ones separated by one second (strictly speaking the maxima are separated by one second but the width of the pulse is negligible compared to their one-second separation). So now we have a device that emits a pulse of negligible width once a second. The question is then, Does it emit a pulse once every second as observed from a different inertial frame? The answer is, of course, yes if we so define it to be. On the other hand, what happens if we compare the time interval as measured by two identical clocks but each in a different inertial frame. To do this we have to make our clock a little more sophisticated. It would clearly be very helpful in constructing the clock if we could make use of invariants, i.e. quantities that will have the same value in all inertial frames. We can no longer take the invariants of the Galilean transformation, because it is precisely that transformation which is suspect. However, we do have from experiment the result that the velocity of electromagnetic waves has the same value in all inertial frames. Thus the value of the velocity of light is one possible candidate for an invariant. Also, in our formal derivation of the Galilean transformation we showed that any length that was at right angles to the relative velocity of the frames was also an invariant. This deduction was based on the principle of relativity alone and should, therefore, continue to hold.

The clock is constructed as follows. We arrange for our one-second pulses to be emitted in the y direction towards a horizontal mirror placed at a distance d along the y axis. The distance d is adjusted so that the reflected pulse just returns to the source as the next pulse is emitted, i.e. one second later. If the velocity of light is c, which has the same value whatever the frame or direction of propagation, then clearly $d = c/2$. The situation is illustrated in Fig. 3.2(a).

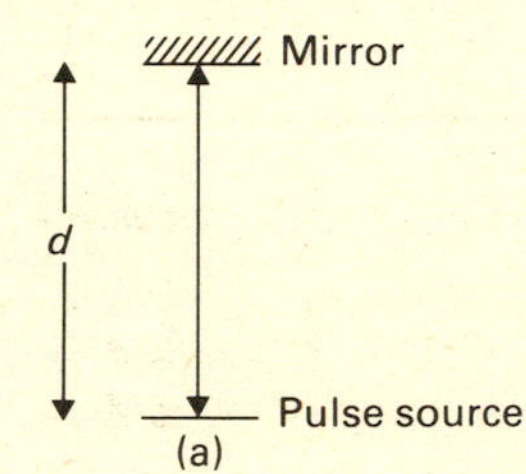

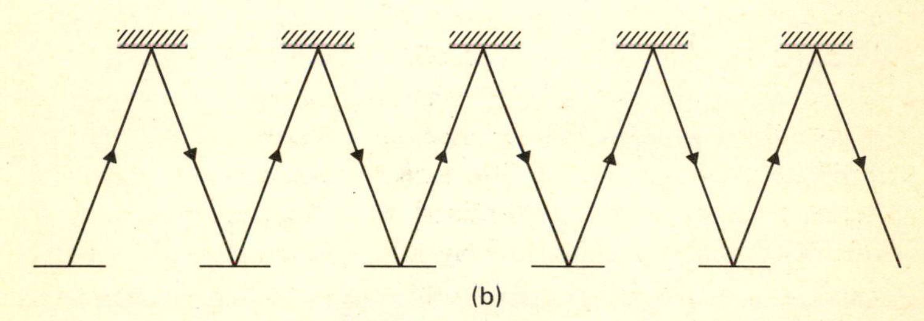

Fig. 3.2 In figure (a) the clock is in the K' frame where the two events A and B occur at the same spatial point. In (b) the clock is viewed from the K frame. In this frame the two events do not occur at the same place and, furthermore, the clock is moving with a velocity v. In both diagrams the 'arrowed' lines show the path of the light pulses.

Suppose that we now place our clock in the frame K' and, further, that there are two events which, when seen by an observer at rest in that frame, occur at the same point in space but are separated in time by n seconds, i.e. n pulses go backwards and forwards between the two events. Now suppose we observe the same two events from the frame K in which we have a clock identical to that placed in the K' frame. Then what is clear immediately is that the two events do not occur at the same place when observed from the K frame (since the K' frame is moving).

That is to say, the spatial coordinates of the two events in the K frame are not the same, unlike the spatial coordinates of the events in the K' frame. In fact, if in the K frame the two events occur at the points x_1 and x_2, then

$$\Delta x = x_2 - x_1 = v\,\Delta t \qquad [3.7]$$

where Δt is the time interval between the two events as measured in the K frame.

We should say a little more about the measurement of Δt. If the events take place at different points on the x axis then we really need two clocks in the frame K, one at x_1 and the other at x_2. This is because we do not wish to move a clock around during the course of the observations. The two identical clocks are synchronized. (This can be done by keeping the clocks at x_1 and x_2 and sending a light signal between the two. Since the velocity of light is a constant the time delay in going from x_1 to x_2 is $(x_2 - x_1)/c$. We do not, of course, really have to know the value of x_1 and x_2 before the observations occur. We can always have an infinite set of identical clocks strung along the x axis, all synchronized with the clock at the origin.)

The behaviour of the clock in K' as seen by the observer at rest in K is shown in Fig. 3.2(b). Now, as velocity of light is the same in all inertial frames, it is immediately obvious that the time between the two events, as observed in the K frame, is going to be greater than the time measured by the clock that is at rest in the K' frame, since the distance is greater. The time observed in the K frame can be calculated as follows. The distance travelled by one pulse between emission and absorption, as observed in the K frame, is, by simple geometry,

$$2[(\Delta x/2n)^2 + d^2]^{1/2} \qquad [3.8]$$

Since the velocity of light is the same in all inertial frames, the time interval, as measured in the K frame, is

$$\Delta t = n(2/c)[(\Delta x/2n)^2 + d^2]^{1/2} \qquad [3.9]$$

Now given that the time $\Delta t'$ between the two events is n seconds in the K' frame and that $d = c/2$, the relationship between the two time intervals is given by

$$\Delta t = (1/c)[(\Delta x)^2 + c^2(\Delta t')^2]^{1/2} \qquad [3.10]$$

This is the remarkable result that Einstein came to after his conversation with his friend Michele Besso:

> An analysis of the concept of time was my solution. Time cannot be absolutely defined, and there is an inseparable relation between time and signal velocity. With this new concept, I could resolve all the difficulties for the first time.

This remarkably simple result already gives us an idea of the new paradigm that Einstein's insight was to open up, namely the indivisibility of space and time. Thus we can rewrite [3.10] as

$$c^2 \Delta t'^2 = c^2 \Delta t^2 - \Delta x^2 = \Delta s^2 \qquad [3.11]$$

Now we could do the same calculation using a different laboratory frame so that the moving frame K' has a different velocity, say v_1. Then in the laboratory frame the separation between the two events would be Δx_1 and the time interval Δt_1, so that

$$c^2 \Delta t'^2 = c^2 \Delta t_1^2 - \Delta x_1^2 = \Delta s^2 \qquad [3.12]$$

The so-called interval Δs is thus an invariant, i.e. it has the same value in all inertial frames. Referring back to Chapter 1, we showed that under a Galilean transformation length was an invariant. However equation [3.12] shows that we must broaden our concept of 'length' and think of it as a four-dimensional concept where the extra dimension in the calculation of length is time. (We make the interval four-dimensional by remembering that the principle of relativity gives us $\Delta y' = \Delta y$ and $\Delta z' = \Delta z$. Thus the four-dimensional invariant interval is $\Delta s^2 = c^2 \Delta t^2 - \Delta x^2 - \Delta y^2 - \Delta z^2$.) We shall return to this concept of a four-dimensional space-time continuum when we derive the full transformation from one inertial frame to another. For the moment we content ourselves with looking at the consequences of equation [3.12].

We have not yet completed our self-appointed task of finding the relationship between the time interval in one frame, where the clock is stationary, and a second frame where the clock is moving with a velocity v. This is now, however, simply done. Substituting equation [3.7] into [3.11] and solving for Δt gives

$$\Delta t = \frac{\Delta t'}{[1-(v/c)^2]^{1/2}} \qquad [3.13]$$

This result is known as time dilation. It says that if two events occur at one place but that they are separated in time by an interval $\Delta t'$, then from a frame in which the first clock is moving with a velocity v the two events will not only be separated in space but also by a time interval, Δt, which is greater than $\Delta t'$ by a factor $1/[1-(v/c)^2]^{1/2}$. The observer in the second frame will conclude that the clock in the frame in which the two events occur at the same place is running slow.

This result or its equivalent (equation [3.11]) is the first new result that we have obtained by insisting that the principle of relativity must apply equally well to electromagnetic theory as to mechanics. Since this result is making a prediction about the real physical world, we should be able to test it experimentally. The next section is devoted to describing two experiments that test the prediction of the time dilation effect.

3.4 Experiments on time dilation

The first experiment to be described was especially performed for pedagogical purposes (see D. H. Frisch and J. H. Smith, *American Journal of Physics,* **31**, no. 5, p. 342, 1963). Nevertheless, it provides a very clear demonstration of the effect we are seeking to verify. Essentially the experiment is very simple; the number of muons (produced by cosmic rays which enter our atmosphere from outer space) in a given time were counted in an observatory on the top of Mount Washington, in the state of New Hampshire. A count for a similar period of time was performed at sea level. In both counts only those muons in a pre-determined band of velocities were counted. After several runs the mean number of muons arriving at the top of the mountain with the pre-selected velocity was 563 ± 10 per hour and the number arriving at the bottom was 408 ± 9 per hour. The pre-selected velocity was $0.9952c$ and the vertical height between the two counting stations was 1907 m.

The two events that we need to consider are a bunch of muons triggering the first counter (at the summit) and the same bunch triggering the second counter (at the foot of the mountain). Now muons are known to have a mean lifetime, when at rest, of $T = 2.21 \pm 0.03 \times 10^{-6}$ s, i.e. if we had N_0 muons at time zero, then after a period of time t, we would expect to have $N_0 \exp(-t/T)$ left. The remainder will have decayed according to the reaction

$$\mu^{\pm} \rightarrow e^{\pm} + \nu + \bar{\nu}$$

The muon can have either a positive or negative charge and thus decays into either an electron (e^-) or positron (e^+) with an associated neutrino (ν) and antineutrino ($\bar{\nu}$). The time taken for any individual muon that survives sufficiently long for the two events to occur, i.e. to trigger both the counters at the top and the bottom of the mountain, is $1907/0.9952c = 6.4 \times 10^{-6}$ s. We would thus expect to count at the bottom of the moutain, given a count of 563 at the top,

$$563 \exp\{-(6.4 \times 10^{-6}/2.2 \times 10^{-6})\} = 31$$

muons, which is nowhere near the experimentally observed value of 408.

Clearly in this calculation the time between events is too long. Let us consider the events in the rest frame of the muon, i.e. a frame which is moving with respect to the Earth's frame at the same velocity as the selected muon velocity. In that frame the two events occur at the same place; if we take the origin of the rest frame to be the muon itself, then both events occur at $x' = 0$.

We shall meet this concept of the rest frame frequently, so it is important to be clear what it means. Imagine that you are sitting on the muon. You will then see the first counter coming towards you. It will hit you at the origin of your coordinate system, to be followed at some later time by the second counter coming towards you. Again it will hit you at the origin of your coordinate system. So far as you are concerned both events—your interaction with the two counters—will have taken place at the same point in space. They will of course be separated in time by some amount $\Delta t'$. The interval between the events, according to equation [3.11], will be given by

$$\Delta s^2 = c^2 \Delta t'^2 \qquad [3.14]$$

In the frame of the Earth the situation is very different, in that the events are separated in both space and time. The spatial separation is the height of the mountain, i.e. the separation between the counters, 1907 m, and the time separation is the time it takes, as measured in the Earth's frame, for the muon to go from the top of Mount Washington to the bottom. This we have already calculated and it turned out to be 6.4×10^{-6} s. Thus in the Earth's reference frame the square of the interval is

$$\Delta s^2 = c^2(6.4 \times 10^{-6})^2 - (1907)^2 = 4.98 \times 10^4 \text{ m}^2$$

Equating this value of Δs^2 with the value given in equation [3.14] above gives $\Delta t' = 7.4 \times 10^{-7}$s. In the rest frame of the muon the number of muons left after such a time interval is

$$563 \exp(-7.4 \times 10^{-7}/2.21 \times 10^{-6}) = 403$$

which is in good agreement with the experimentally observed number of 408 ± 9.

The second example of time dilation is very similar to the first. This is to some extent unavoidable, as equation [3.13] shows that the time dilation effect is negligible unless $v/c \sim 1$. To accelerate particles to such velocities means that for practical reasons we are restricted to particles of very small mass, and in particular to the so-called elementary particles. (There are exceptions to the rule. For example, an atomic clock has been flown round the earth in a jet plane and compared with a clock that remained on Earth. Because the orbit is circular the clock is accelerating and the experiment requires a more complicated analysis than that given here. See J. C. Hafele and R. E. Keating, *Science*, vol. 177, pp. 166, 168, 1972.)

Our second example, therefore, is a measurement of the lifetime of charged pions, mass = 140 MeV = 2.2×10^{-11} kg, at high velocities (see A. J. Greenberg *et al.*, *Physical Review Letters*, vol. 23, no. 21, p. 1267, 1969). It is customary to refer to the masses of elementary particles in energy units, and in particular in electron-volts. What is really being referred to is the rest energy, mc^2 (see section 5.4.). Thus 1 eV is the energy acquired by an electron accelerated through one volt, which is equal to 1.602×10^{-19} J. So the mass equivalent to 1 eV is $1.602 \times 10^{-19}/(3 \times 10^8)^2 = 1.78 \times 10^{-36}$ kg.

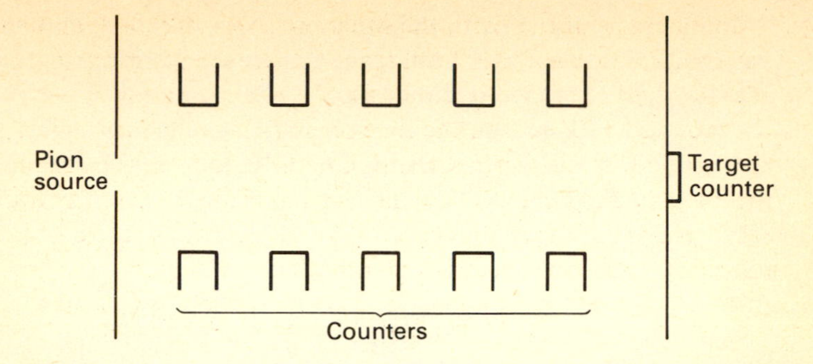

Fig. 3.3 Schematic experimental set up for the measurement of the pion lifetime when the pion is moving with a velocity close to the velocity of light.

A schematic illustration of the experimental set up is shown in Fig. 3.3. It consists of a pion source and a target counter towards which the pions are directed; along the flight path of the pions are counters spaced roughly at 6-foot intervals. In the laboratory frame the number of pions per unit length when emitted from the source is N_0. After a time t the number of remaining pions is

$$N(t) = N_0 \exp(-t/T)$$

where T is the lifetime of the pions. For pions with a selected velocity v, the distance that the pions have travelled from the source $x = vt$. Thus the number of pions per unit length at a distance x from the source is

$$N(x) = N_0 \exp(-x/vT)$$

The ratio of the number of counts at a distance x from the source to the number at the source is

$$R(x) = N(x)/N_0 = \exp(-x/vT)$$

If the logarithm of this ratio for each of the counters is plotted versus their distance from the source, a straight line should result whose slope is $-1/vT$. In the experiment v, as measured by time of flight, was found to be $0.91c$ and the reciprocal of the slope of the graph -17.1 m. Hence

$$T = 17.1/0.91c = 6.26 \times 10^{-8}\ \text{s}$$

The lifetime of the pion when at rest has been measured and its value is 2.604×10^{-4} s. If we apply the time dilation formula (equation [3.13]) directly, remembering that the rest frame of the pion is the frame where the 'clock' is at rest and the laboratory frame the one where it is moving, then the time dilation factor $[1-(v/c)^2]^{-1/2}$ is

$$6.26 \times 10^{-8}/2.60 \times 10^{-8} = 2.41$$

The value of the velocity corresponding to this value is $0.91c$, which, within the limits of experimental error, agrees exactly with the measured value.

We chose to analyse two similar experiments by two different methods, the first by using the invariance of the interval and the second by a straight application of the time dilation formula. Both of these methods are equivalent, since the second follows from the first by the use of equation [3.7]. Which method is employed is simply a matter of personal preference. The advantage of the time-dilation formula is that it essentially reduces to putting numbers in a formula. The disadvantage is that it is easy to confuse in which frame time is dilated. The advantage of the invariance of the interval method is that the latter problem is avoided. The disadvantage is that it actually means deriving the time dilation formula every time.

3.5 The Lorentz-FitzGerald contraction

Both the experiments described in the last section lead to an unexplained inconsistency. This is easiest to see in the muon experiment.

Consider the situation in the rest frame of the muon. The question remains: In this frame what do we take for the velocity of approach of the counters? The formal derivation of the Galilean transformation given at the beginning of this chapter was quite explicit; the velocity, since we are in the rest frame of the muon, is numerically equal to the velocity of the muon in the Earth or laboratory frame. Remembering that this part of the derivation is a part that we decided to keep, the distance between the two events is therefore $0.9952c \times 7.4 \times 10^{-7} = 220$ m. Now since, in the Earth's frame, the distance between the two events is the height of the mountain, in the muon's frame the mountain appears to have shrunk to approximately one-tenth of the height as seen by a stationary observer on Earth! If we had done the calculation in symbols rather than numbers we would have found that if h is the height of the mountain in the Earth's frame and h' the height in the muon's frame, then

$$h' = h[1-(v/c)^2]^{1/2} \tag{3.15}$$

It would appear therefore that it is not only time that differs from reference frame to reference frame but also length.

Actually this isn't surprising if we refer back to equation [3.12]; since Δs^2 has the same value in all inertial frames and Δt can vary from frame to frame, it follows immediately that Δx also has to vary from frame to frame. The result (equation [3.15]) is the Lorentz-FitzGerald contraction mentioned in the introduction to Chapter 1.

We have to be a little careful in interpreting this result. What we actually measure in the muon frame is the time between events and the velocity with which the two counters approach. From these two observations we deduce the distance between the two events. This is a rather roundabout way of deducing the Lorentz-FitzGerald contraction. The question we want to answer is, If the length of a rod is l' when it is at rest in the K' frame, what is its length when measured in the K frame? To answer this question experimentally we would have to observe the positions of the two ends of the rod on the x axis, the observations to be made at the same time, since the rod is moving in this frame. We will be able to predict the result of this experiment when we have obtained the correct transformation between inertial frames. This will be done in Chapter 4. The answer will be found to be identical to equation [3.15].

3.6 Causality

Before going on to derive the correct transformation for the space-time coordinates, there is one other very important result we can derive. Suppose we consider the interval between two events where one event occurs at the origin of both space and time; then we can write

$$s^2 = (ct)^2 - x^2 \qquad [3.16]$$

The necessary condition for two events to be causally related is that the event being caused shall occur at a later time than the event which causes it. Two comments on this definition. First, merely because it is a necessary condition, it does not follow that if one event is later in time than another it has to be caused by the first. Second, if two events are causally related in one frame, to comply with the principle of relativity they must be causally related in all inertial frames. If this were not so then we could distinguish between inertial frames by whether two physical events were causally related or not.

Now let us assume that we can find a reference frame where two causally connected events both occur at the spatial origin. Then for this frame the s of equation [3.16] can be written

$$s^2 = (c\tau)^2 \qquad [3.17]$$

where τ is known as the proper time. The proper time has a particular significance. In the time-dilation experiments described in the last section, the time measured in the rest frame of the muon is the proper time, for in that frame two causally connected events – the coincidence of the muon with the two counters – both occur at the origin. In general the time as measured in the rest frame of a moving object is the proper time.

It follows from equation [3.17] that the interval between causally connected events is always positive. Since s has the same value in all inertial frames, it follows that in an inertial frame in which two events are separated in space by x and in time by t that

$$(ct)^2 > x^2 \qquad [3.18]$$

Such an interval is known as a time-like interval.

If $s^2 < 0$ it follows that

$$(ct)^2 < x^2 \qquad [3.19]$$

In this case the interval is said to be space-like and if we put $\sigma^2 = -s^2$, σ is known as the proper distance. Thus any two events that have a space-like interval cannot be causally connected. Finally, if

$$(ct)^2 = x^2 \qquad [3.20]$$

the interval is called light-like.

The concept of the interval can, as mentioned above, be extended to three spatial dimensions and the obvious generalizations of time-, space- and light-like intervals can be made. Figure 3.4 shows how the two-spatial, one-time dimension continuum is divided by the different types of intervals. The two-spatial dimensional equivalent of equation [3.20] is

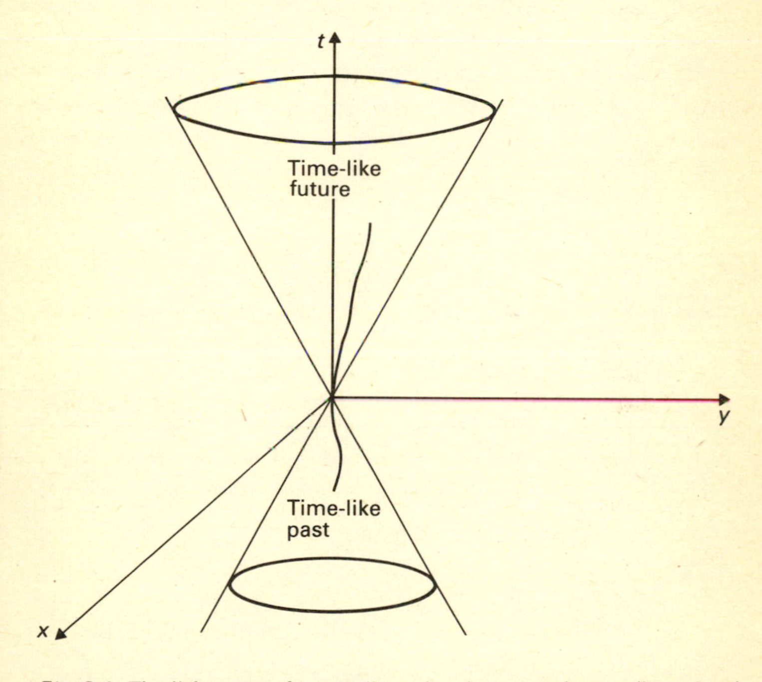

Fig. 3.4 The light cone of a two-dimensional space and a one-dimensional time continuum. The irregular line is the world line of a particle that has passed through the origin. For an event at the origin, any other event that is causally connected to it must lie within the light cone.

$$(ct)^2 = x^2 + y^2 \qquad [3.21]$$

In the three-dimensional space-time continuum this is the equation of a cone. Inside the cone the interval is time-like, i.e.

$$(ct)^2 > x^2 + y^2 \qquad [3.22]$$

Outside the cone the interval is space-like and the opposite equality to equation [3.22] holds. The cone is known as the light cone and, remembering that this interval is defined for one event at the origin, it follows that the second event must be in or on the upper half-cone if it is to be caused by the event at the origin. Similarly, the second event could have caused the first (first and second clearly do not refer to a time ordering) if it lies in the bottom half of the light cone. For this reason the volume of the continuum bounded by the upper half of the light cone is known as the time-like future, whilst the volume bounded by the lower half is known as the time-like past. Since any event which lies outside the light cone has a space-like interval between it and the origin, it follows that there can be no causal connection between the two events.

To pursue the concept of causality further we must consider what we mean by two events being causally connected. So far we have merely given a necessary condition. A second condition is clearly that for event A to have caused event B, information must have flowed from A to B. If that had not occurred then it would mean that event B would have occurred without any knowledge of event A having happened. That could not be interpreted as event A causing event B. For the case where A causes B, let v_{I} be the velocity with which information is conveyed from A to B. Then the minimum time which can elapse between events A and B is

$$t = x/v_{\mathrm{I}}$$

where x is the spatial separation between A and B. For the case where the time separation is a minimum the interval is given by

$$s^2 = (ct)^2 - x^2 = t^2(c^2 - v_{\mathrm{I}}^2) \qquad [3.23]$$

but since the events are causally related, i.e. $s^2 \geqslant 0$, it follows

$$v_{\mathrm{I}} \leqslant c.$$

This is a very important and far-reaching result; it says that it is impossible to convey information at a velocity which is greater than the velocity of light. Why it is important will become apparent later, but if the reader refers back to our reformulated laws of motion in Chapter 2 he/she will begin to see the consequences upon the foundations of classical mechanics.

At first glance it may seem that this result has to be wrong, in the sense that it is not difficult to find apparent exceptions. For example, suppose that we have two spaceships at a distance of 6×10^8 m from Earth and separated by some 10^5 m. Again, suppose that we have on Earth a very powerful laser capable of focusing a beam on either of the two ships. We arrange that the laser is rotated at an angular speed of one radian per second so that the spot of light travels from one spaceship to the other. Now the speed at which the spot travels is $1 \times 6 \times 10^8$ m s^{-1}, which means that the spot travels the distance of 10^5 at exactly twice the speed of light! Further we could arrange that the spaceships start their engines when the spot reaches their particular craft. Has information been conveyed at a speed that is greater than the velocity of light? The answer according to Einstein is clearly no, and, equally clearly, that is correct. The result (equation [3.24]) does not say that we cannot have velocities greater than the velocity of light but only that information cannot be conveyed at speeds greater than that velocity. So there is nothing inviolate about the spot moving across space at a velocity of $2c$. Nor does the firing of the engines imply that information has been conveyed at that speed. The firing of the engines is in response to information that has been relayed to each ship earlier; the spot reaching each of them is merely acting as a clock, telling them *independently* when to start their engines. The engine of each of them would be fired irrespective of whether the other was started or not.

With this discursion into the realms of science fiction we end this chapter and in the next obtain the correct coordinate transformation in going from one inertial frame to another.

Chapter 4
The Lorentz transformation

4.1 Introduction

We started the last chapter by formally deriving the Galilean transformation in an attempt to try and identify any wrong assumptions in the argument. We came to the conclusion that the only part of the argument that we could throw away was that time was the same in all frames of reference. This means that in order to find the correct transformation we have to find an alternative assumption to this one. In fact we have two alternative choices; either we can take the velocity of light or the interval as an invariant. Clearly there is no conflict between these two statements, since we derived the second from the first. Which we take is a matter of personal choice, so let us take the invariance of the velocity of light, simply on the grounds that it is the experimentally observed invariant.

4.2 The Lorentz transformation

Using the principle of relativity and the definition of velocity, we arrived in Chapter 2 at the relations [3.1] and [3.6], i.e.

$$y' = y \qquad z' = z \qquad [4.1]$$

$$x' = a(x - vt) \qquad t' = ex + at \qquad [4.2]$$

For the Galilean transformation we made the further assumption that $t' = t$ to identify e and a. To make use of our new assumption, i.e. the constancy of the velocity of light, we first derive the equations for the transformation of the velocities. Thus

$$u'_x = \frac{dx'}{dt'} = \frac{a(dx - vdt)}{edx + adt} = \frac{a(u_x - v)}{eu_x + a} \qquad [4.3]$$

$$u'_y = \frac{u_y}{eu_x + a} \qquad u'_z = \frac{u_z}{eu_x + a} \qquad [4.4]$$

Consider first a signal which is propagating along the x axis of the K frame with the velocity of light. Since the velocity of light is an invariant, it follows that the velocity of the signal in K' is also c. Thus from [4.3]

$$c = \frac{a(c - v)}{ec + a}$$

or

$$ec^2 + av = 0 \qquad [4.5]$$

Suppose now that the direction of the signal propagation in the K frame is changed so that it is now propagating along the y axis. Then, according to equations [4.3] and [4.4], there will be both an x' and a y' component of the signal velocity in the K' frame. These components will be

$$u'_x = -v \qquad u'_y = c/a$$

The invariance of the velocity of light then tells us that it is $(u'^2_x + u'^2_y)$ which should be equal to c. Hence

$$c^2 = v^2 + (c/a)^2$$

or

$$a = \frac{1}{[1 - (v/c)^2]^{1/2}}$$

Substituting in [4.5] gives for e

$$e = -\frac{1}{[1 - (v/c)^2]^{1/2}} \frac{v}{c^2}$$

Using these values of a and e in [4.2] gives the Lorentz transformation

$$\begin{aligned} y' &= y \qquad\qquad z' = z \\ x' &= \gamma(x - vt) \\ t' &= \gamma(-vx/c^2 + t) \\ \gamma &= 1/[1-(v/c)^2]^{1/2} \end{aligned} \tag{4.6}$$

The inverse transformation is obtained by interchanging the primed and unprimed quantities and changing the sign of v, i.e.

$$\begin{aligned} y &= y' \qquad\qquad z = z' \\ x &= \gamma(x' + vt') \\ t &= \gamma(vx'/c^2 + t') \end{aligned} \tag{4.7}$$

These then are the equations that are to replace the Galilean transformation (equations [1.6]). Before discussing their significance, we can complete the transformation by obtaining the equations for the transformed velocities. This is easily done by substituting the values of a and e we have just obtained in equations [4.3] and [4.4], thus

$$\begin{aligned} u'_x &= \frac{u_x - v}{(1 - u_x v/c^2)} \\ u'_y &= \frac{u_y}{\gamma(1 - u_x v/c^2)} \\ u'_z &= \frac{u_z}{\gamma(1 - u_x v/c^2)} \end{aligned} \tag{4.8}$$

These transformation equations for the velocities are considerably more complicated than the corresponding ones for the Galilean transformation (equation [1.7]). It is not surprising, therefore, that the equations for transforming the acceleration are even more complicated. The procedure is, however, straightforward and is left as an exercise for the reader (see exercise 2, Chapter 4, Appendix 3). The result is

$$a'_x = \frac{a_x}{\gamma^3(1 - u_x v/c^2)^3}$$

$$a'_y = \frac{a_y}{\gamma^2(1-u_x v/c^2)^2} + \frac{a_x(u_y v/c^2)}{\gamma^2(1-u_x v/c)^3} \qquad [4.9]$$

$$a'_z = \frac{a_z}{\gamma^2(1-u_x v/c^2)^2} + \frac{a_x(u_z v/c^2)}{\gamma^2(1-u_z v/c^2)^3}$$

It is interesting to note that, although the set of equations [4.6] is known as the Lorentz transformation, it was not Lorentz who first wrote them down. In his theory of the electron, Lorentz had used a transformation which was identical to that obtained by expanding equations [4.6] in powers of v/c and retaining only the first two terms. The first term in such an expansion is obtained by putting $v/c = 0$, and the result is the Galilean transformation (equation [1.6]). This result is comforting, in that at velocities much less than the velocity of light we might expect the Galilean transformation to hold. According to Whittaker (see bibliography, Appendix 1) one of the first people to write down the full Lorentz transformation was Voight, in a paper on the theory of vibratory motions. Nevertheless it is fitting that Lorentz, with his great contributions to physics, should have this very important set of equations named after him.

The Lorentz transformation gives another illustration of a property that we have already met. In the discussion of the Galilean transformation of electric and magnetic fields we found, for example, that what in one frame of reference was an electric field became in another frame of reference both an electric and a magnetic field. The Lorentz transformation is another example of the mixing of different quantities when going to a different inertial frame. Up until now we have treated time and space as though they were separate and different concepts. Equations [4.6] show that space and time become mixed up in going from one inertial frame to another. In the K' frame the x' and t' 'coordinates' are admixtures of the x and t 'coordinates' in the K frame. Thus we have to blur the distinction between space and time, and really we should only talk about the space-time continuum. From a practical point of view, however, equations [4.6] tell us how to relate the four-dimensional coordinates in one inertial frame to the four-dimensional coordinates of the same event in a different inertial frame which is moving along the x axis of the first frame with velocity v.

This discussion of the transformation from one inertial frame to another began because Maxwell's equations were not form invariant under the Galilean transformation. To be consistent we ought to check whether under the Lorentz transformation they are form invariant. However rather than do that now we will pursue the consequences of the Lorentz transformation and return to the subject of electromagnetism in Chapter 7.

Before looking at some of the consequences on mechanics in Chapter 5 we must, as we promised, derive the Lorentz-FitzGerald contraction.

4.3 The Lorentz-FitzGerald contraction

We have already described in Chapter 3 the relationship between the length of a rod when it is at rest and when it is moving with respect to an observer. The problem with that derivation was that it was accomplished in a rather roundabout fashion. The Lorentz transformation (equations [4.6]) enables us to obtain the result in a straightforward manner. We want to measure the length of a stick when it is at rest in the K frame. We do this by placing it along the x axis and noting the values of x where the two ends of the rod lie. Let these two values be x_1 and x_2, where $x_2 > x_1$. Then the length of the stick is

$$l = x_2 - x_1$$

An observer in the K' frame wishing to make the same measurement has to observe the ends of the rod and note where they come on his x' axis. However in his case he must ensure that the observations are made at the same time, since in his frame of reference the stick is moving. Let the two values of x' be x'_1 and x'_2, where $x'_2 > x'_1$, so that the length in the K' frame is

$$l' = x'_2 - x'_1$$

and the time at which the measurements are made t'. Then according to the inverse Lorentz transformation (equations [4.7])

$$x_1 = \gamma(x_1' + vt')$$
$$x_2 = \gamma(x_2' + vt')$$

Hence

$$l = x_2 - x_1 = \gamma(x_2' - x_1')$$

or $$l' = l[1 - (v/c)^2]^{1/2} \quad [4.10]$$

which apart from a change in notation is the same as the result in equation [3.15].

The problem with the Lorentz-FitzGerald contraction is that it is very difficult to verify experimentally. The difference between the two lengths is of the order $(v/c)^2$ and, as we have remarked in the case of time dilation, to get values of v that would make this difference significant we have to have elementary particles because their small mass enables them to be accelerated to high velocities. However for elementary particles there is a fundamental difficulty in defining the length when we are down to the scale of these particles. To describe the particles we have to use quantum mechanics, where the 'particle' is represented by a wave packet, its length no longer being unambiguously defined. To have a well-defined length we need a macroscopic body and then we have the problem of accelerating that body to speeds comparable to the speed of light. One possibility is to use the Earth as a reference frame and one of the other planets as the object to be measured. For example, Venus rotates round the Sun at a rate that is faster than the Earth, so in one aspect the Sun is between the Earth and Venus, whilst in the other aspect Venus is between the Earth and the Sun. As an order of magnitude the Earth's velocity round the Sun is 30 km s^{-1} and that of Venus 36 km s^{-1}. Thus when the Earth and Venus are adjacent their relative velocity is 6 km s^{-1}, which is approximately zero compared to their relative velocity in the other aspect of 66 km s^{-1}. Thus we can take $v/c \simeq 6.6 \times 10^4 / 3 \times 10^8 = 2.2 \times 10^{-4}$ and hence

$$l' = l(1 - 4.8 \times 10^{-8})^{1/2} \simeq l(1 - 4.8 \times 10^{-8}/2)$$

or $$(l - l')/l = 2.4 \times 10^{-8}$$

Thus to detect the Lorentz-FitzGerald contraction we would

have to measure the diameter of Venus from the Earth with an accuracy that is better than one part in 10^8. The diameter of Venus is approximately 1.24×10^7 m, so that we would have to have an instrument that would be able to detect differences of less than a tenth of a metre. Apart from the fact that this is considerably less than the fluctuation in the diameter of Venus, there would be other problems in the experiment. The measurement would have to be done using radiation and to detect differences of order of centimetres we would have to use radiation of wavelength of that order. Until the recent advent of space probes, nothing was known about the surface of Venus, which suggests that radiation from the surface was being totally absorbed by the cloud layer. (At the time of writing the latest Soviet spacecraft, Veneras 15 and 16, are expected to orbit Venus shortly. Part of their mission is to map the surface of the planet using radar.) Thus the choice of Venus was not a very good one for verifying the Lorentz-FitzGerald contraction. Exercise 11 (Chapter 4, Appendix 3) asks you to repeat the calculation for some other planets to see if they are more suitable candidates.

That concludes our discussion of the Lorentz transformation. In the next chapter we will go on to look at its implications for the study of mechanics.

Chapter 5
The conservation of energy and momentum

5.1 Introduction

We have now arrived at the point where we need to take stock. In the last chapter we derived the Lorentz transformation, which we claimed was the correct transformation of the coordinates of an event from one inertial frame to another. The results were quite different from the corresponding ones for the Galilean transformation. On the other hand, the Galilean transformation did leave Newton's equations form invariant. The question then remains, how do these equations transform under the Lorentz transformation? Are they form invariant? If they are not, we seem to be in the same difficulty that led us to derive the new transformation, namely that half of nineteenth-century physics does not satisfy the principle of relativity.

The behaviour of the equations of motion under the Lorentz transformation is easily found. In Chapter 2 one of our reformulated laws (equation [2.3]) was that the ratio of the accelerations of two interacting particles is a constant. If we assume that this law holds in the frame K' and also, just to simplify matters, that the acceleration is directed solely in the x direction, i.e.

$$a'_{1x}/a'_{2x} = k_{12}$$

then using equation [4.9] the law in frame K becomes

$$\frac{a_{1x}}{a_{2x}} = \frac{k_{12}(1-u_{1x}v/c)^3}{(1-u_{2x}v/c)^3}$$

This is a result which is manifestly not form invariant, since both

u_{1x} and u_{2x} appear on the right-hand side. So it seems we really do have a problem. However a little consideration shows that the situation need not be as bad as it appears. First, it has to be remembered that Newton's equations, or equivalently our reformulated laws, have stood the test of time, from the point of view of experiment, very well indeed. The reason for this is not hard to find. We found in discussing the Lorentz-FitzGerald contraction in the last chapter that it was very difficult to observe this effect because it really only appeared as a measurable quantity when the relative velocity of the inertial frames was comparable to the velocity of light. Until well into this century no body or particle had been observed at anywhere near this velocity. It follows, therefore, that one possibility, and at the moment it is only a possibility, is that Newton's equations are a good approximation to the true equations of motion when the velocities of the particles are small compared to the velocity of light. Support for this possibility is given by the fact that we know that for the case of low velocities the Lorentz transformation reduces to the Galilean transformation. Thus, if we are going to try and reformulate the laws of mechanics, then as a guide we would expect them to reduce to Newton's equations in the limit $v/c \to 0$.

5.2 Particle dynamics

Rather than try to write down the general equations of motion right from the start, we start by being less ambitious. Many problems in Newtonian mechanics can be solved without solving the full equations of motion but instead using the equations for the conservation of momentum and energy. Such problems require us to be given data about the final states of the system which we would not need if we were solving the full dynamic equations; this is frequently the case in experiments involving the collision between two particles. We shall postpone formulating the Lorentz invariant form of the equations of motion to Chapter 7. The reason that we have chosen to deal with the more restricted problem first is not hard to see; the equations for the transformation of the accelerations (equations [4.9]) are really rather cumbersome. The equations [4.6] for the transformation of the velocities are not as simple as those for the Galilean transformation but they are a great deal simpler than those for the acceleration. This suggests that it might be simpler to try and formulate, in the first instance, a law which involves the velocities. Clearly we cannot formulate the full dynamic equations in this way since, in the Newtonian limit, acceleration must appear.

Before going on to look at what can be done towards solving the restricted problem, there is one fundamental difficulty which should be mentioned in the formulation of any dynamic laws. In our formulation of Newton's laws we had the proposition that there exists a law of force which, at the most, can only depend upon the difference between the positions and the difference between the velocities of the particles (equation [2.11]). Suppose initially that both particles are at rest. Then, if one of the particles moves, the force acting on the other will instantaneously change. Since force is defined as mass times acceleration, the second particle will start to accelerate. In other words we have 'action at a distance', a concept which disturbed Newton:

> That one body may act upon another at a distance through a vacuum, without the mediation of anything else, by which their action and force may be conveyed from one to another, is to me so great an absurdity, that I believe no man, who has in philosophical matters a competent faculty of thinking, can ever fall into it. (Quoted in *Understanding Space and Time*, Block 1, S354, Open University, Milton Keynes.)

This effect will occur however far apart the particles are, provided the force is not identically zero beyond a certain separation of the particles. Thus particle 2 'knows' that particle 1 has moved and, further, that this information has been conveyed with an infinite velocity.

Now one of the results of Chapter 3 was that there existed an upper bound, the velocity of light, for the velocity by which information is conveyed. It follows, therefore, that the concept of a law of force cannot in general hold, at least not in the Newtonian sense. It must, however, be emphasized that the concept of two particles interacting through a law of force will be a very good approximation to reality in the case where the velocities of the particles are small compared to that of light.

The above argument does not necessarily affect the definitions of the electric and magnetic fields, for in those cases the force acting on the particle depends only on the field at the position of the particle. Thus in this case action at a distance will not occur.

Having mentioned the difficulties in formulating a Lorentz invariant form of the equations of motion, we now return to the simpler task of formulating relativistically invariant forms for the laws of conservation of energy and momentum. As we have emphasized, our reformulated Newton's laws are expected, on the grounds of experiment, to hold in the low-velocity limit. Thus equations [2.3], [2.4] and [2.5] will still hold, provided that adjacent to each equation we put the condition 'in the limit as $v/c \to 0$'. Now we can reproduce all the arguments that led us from these equations to the definition of mass, all the while remembering that the definition only holds in the above limit. For this reason what we have previously referred to as 'mass' we shall in future refer to as the 'rest mass'. Initially we will use the symbol m_0 to differentiate it from the Newtonian mass, which is defined at all velocities. We should note, however, that numerically $m_0 = m$. In the Newtonian regime, once mass had been defined, it was possible to derive from the basic laws the law of conservation of momentum, expressed in equation [2.7]. We should first test if that equation is form invariant under the Lorentz transformation.

Applying the equations for transforming the velocities (equations [4.8]) shows immediately that the law does not have the same form in all inertial frames. We now have two choices; either we abandon the possibility of having a law of conservation of momentum or, and this appears to be the only other possibility, the Newtonian expression for the momentum of the particle has to be modified.

Under the Galilean transformation the velocities transform according to equations [1.7], which means that the components of the momentum transform like

$$p'_x = p_x - vm_0 \qquad p'_y = p_y \qquad p'_z = p_z \qquad p'_t = p_t$$

where $p_t = m_0$. That is to say, if we regard the three components of $\mathbf{p}$ (p_x, p_y and p_z) along with the 'time-like' component p_t, as a four-dimensional vector, then the four components transform under the Galilean transformation in exactly the same manner as the space and time coordinates. Further, suppose we write down the law of conservation of momentum in the frame K':

$$\mathbf{p}' = \mathbf{i}p'_x + \mathbf{j}p'_y + \mathbf{k}p'_z = \mathbf{C}$$

(where $\mathbf{C}$ is a constant vector) then using the above transformation, the law in the K frame becomes

$$\mathbf{p} = \mathbf{C} + m_0\mathbf{v}$$

Since both m_0 and $\mathbf{v}$ are constants, the form of the law is the same. This suggests that we ought to be looking for a four-dimensional vector that transforms like the Lorentz transformation. Such vectors are known as four-vectors. So, before pursuing the problem of the correct expression for the momentum further, we digress to investigate the properties of such vectors.

5.3 Four-vectors

A four-vector is defined as one whose four components – say F_x, F_y, F_z, F_t – transform like the Lorentz transformation, i.e. their corresponding values in the K' frame are

$$\begin{aligned} F'_x &= \gamma(F_x - vF_t) \\ F'_y &= F_y \qquad\qquad F'_z = F_z \\ F'_t &= \gamma(-vF_x/c^2 + F_t) \qquad\qquad [5.1] \\ \gamma &= \frac{1}{[1-(v/c)^2]^{1/2}} \end{aligned}$$

The inverse transformation has exactly the same form as the inverse transformation for x, y, z, t. Examples of four-vectors are the space-time coordinates themselves, the difference between any two sets of coordinates, i.e. $\Delta x = x_2 - x_1$, $\Delta y = y_2 - y_1$, $\Delta z = z_2 - z_1$, $\Delta t = t_2 - t_1$, and either of these two vectors multiplied or divided by a constant scalar. Other examples will appear later in the book. The second example is easily proved by writing down the Lorentz transformation for each of the sets of coordinates and then taking the difference. The last example follows immediately. What are the properties of four-vectors?

For the four-vector $\vec{\mathbf{r}} = (\mathbf{r}, t)$ (we shall denote four-vectors in bold-faced type, as for three-vectors, but distinguish them by putting an arrow above) an invariant under the Lorentz transformation was the square of the interval

$$s^2 = c^2t^2 - r^2$$

This suggests that if we define the 'length' $\vec{F}$ of a four-vector $\vec{\mathbf{F}}$ by the equation

$$\vec{F}^2 = c^2F_t^2 - F_x^2 - F_y^2 - F_z^2 = c^2F_t^2 - F^2 \qquad [5.2]$$

then perhaps $\vec{F}^2$ would be an invariant, i.e. have the same value in all inertial frames. In fact there is a more general result than this. If we define the scalar product of two four-vectors $\vec{\mathbf{F}}$ and $\vec{\mathbf{G}}$ by

$$\vec{\mathbf{F}} \cdot \vec{\mathbf{G}} = c^2F_tG_t - F_xG_x - F_yG_y - F_zG_z$$

$$= c^2F_tG_t - \mathbf{F} \cdot \mathbf{G} \qquad [5.3]$$

then exercise 1, Chapter 5, Appendix 3 asks you to show that this quantity has the same value in all inertial frames. The case $\vec{\mathbf{F}} = \vec{\mathbf{G}}$ shows that the 'length' of the four-vector is an invariant.

In the previous section we showed that the Newtonian law of conservation of momentum, which has the form momentum equals a constant, was form invariant under a Galilean transformation but not under a Lorentz transformation. The question is, then, would a law of the form a four-vector equals a constant be form invariant under a Lorentz transformation? Suppose we have a law of the form

$$\vec{\mathbf{F}} = \vec{\mathbf{C}} \qquad [5.4]$$

where $\vec{\mathbf{C}}$ is a constant four-vector. Then in terms of the components of $\vec{\mathbf{F}}$, the law can be written,

$$F_x = C_x \qquad F_y = C_y \qquad F_z = C_z \qquad F_t = C_t$$

Applying the Lorentz transformation (equations [5.1]) to these equations gives the new law in the frame K', i.e.

$$F'_x = \gamma(F_x - vF_t) = \gamma(C_x - vC_t) = C'_x$$

$$F'_y = F_y = C_y = C'_y \qquad F'_z = F_z = C_z = C'_z \qquad [5.5]$$

$$F'_t = \gamma(-vF_x/c^2 + F_t) = \gamma(-vC_x/c^2 + C_t) = C'_t$$

Since, as for the Galilean transformation, v, the relative velocity of the frames, is a constant, the law in the frame K' has the same form as in K. In four-vector notation equations [5.5] are written

$$\vec{\mathbf{F}}' = \vec{\mathbf{C}}' \qquad [5.6]$$

To summarize this digression, a four-vector is analogous to the three-vector of Euclidean space; it has the property that its 'length' is an invariant under a Lorentz transformation. Further, if we have a physical law of the form a four-vector equals a constant, then that law is form invariant under the Lorentz transformation.

5.4 The conservation law

From the discussion in the last section it is apparent that if we can find the appropriate four-vector, then that vector equals a constant and would be a candidate for the law of conservation of momentum. Reference back to equation [4.8] for the transformation of the velocities shows that they do not transform like the Lorentz transformation, i.e. the transformation equations are not those of equations [5.1].

However, let us consider how we obtain the velocity of a particle. We first write down the time it takes, Δt, to pass two markers separated by a distance $\Delta\mathbf{r}$. This distance will have three components and if we include Δt then, as shown in the previous section, we have the four components of a four-vector. Now in calculating the velocity we divide these components of the change in position by Δt and, as we have already remarked, the subsequent expressions do not form a four-vector. In Chapter 3 we discussed the constancy of the velocity of light in all inertial frames and showed that it led to time being different in different inertial frames. There was one particular frame, however, for which the time interval was of particular interest; this was the frame where two causally connected events occurred at the same point in space. For such a frame the time interval was known as the proper time interval. In the examples that we gave in Chapter 4 on time dilation, that particular frame for a particle was its rest frame. Thus in measuring the velocity of the particle the two events are the positions of two markers, $\Delta\mathbf{r}$ apart. In the rest frame of the particle these two events occur at the same point and are therefore separated by the proper time interval. The time interval in the original frame is related by the time-dilation formula (equation [3.13]), i.e.

$$\Delta t = \frac{\Delta\tau}{[1-(u/c)^2]^{1/2}} \qquad [5.7]$$

where u is the velocity of the particle as measured in the original frame. Since the interval is an invariant and for causally connected events is equal to $c\Delta\tau$, the proper time interval will also be an invariant, remembering that c has the same value in all inertial frames. Thus if we divide the four-vector $(\Delta\mathbf{r}, \Delta t)$ by $\Delta\tau$ rather than Δt, we will still have a four-vector. Finally, on the grounds that at the low velocity limit we need to produce the Newtonian expression for the momentum, we multiply by the rest mass m_0. Thus we define the Lorentz invariant form of the four components of momentum as

$$p_x = m_0\frac{\Delta x}{\Delta\tau} \qquad p_y = m_0\frac{\Delta y}{\Delta\tau}$$
$$p_z = m_0\frac{\Delta z}{\Delta\tau} \qquad p_t = m_0\frac{\Delta t}{\Delta\tau} \qquad [5.8]$$

If we use equation [5.7], then these expressions can be written

$$p_x = \frac{m_0 u_x}{[1-(u/c)^2]^{1/2}} \qquad p_y = \frac{m_0 u_y}{[1-(u/c)^2]^{1/2}}$$
$$p_z = \frac{m_0 u_z}{[1-(u/c)^2]^{1/2}} \qquad p_t = \frac{m_0}{[1-(u/c)^2]^{1/2}} \qquad [5.9]$$

Some authors summarize these equations as

$$\mathbf{p} = m\mathbf{u} \qquad p_t = m$$

where

$$m = m_0/(1-u^2/c^2)^{1/2}$$

i.e. the expression for the momentum is formally identical to the Newtonian expression but the mass is velocity dependent. Appealing though this is, it disguises the fact that the difference between equations [5.9] and the Newtonian expressions comes from dividing by $\Delta\tau$ rather than Δt. This has nothing to do with our definition of mass.

These then are our postulated expressions for the four-momentum. In fact we have a problem, since we have four components of momentum rather than three and, further, the fourth component is not immediately identifiable as a physical quantity. However we have always the check that in the low velocity limit, i.e. $u/c \to 0$, we should regain the Newtonian expressions. This should help us to identify the fourth component. But before we do that we will check the values of the first three components in that limit. This is very easily done by putting $(u/c) = 0$ in the first three of equations [5.9]. It is seen immediately that these expressions reduce to the Newtonian ones. For the fourth component we expand in powers of (u/c), using the binomial expansion:

$$p_t = \frac{m_0}{[1-(u/c)^2]^{1/2}} = m_0 + \frac{m_0 u^2}{2c^2} + \ldots \qquad [5.10]$$

The second term on the right-hand side of this equation is the familiar Newtonian expression for the kinetic energy of the particle divided by c^2. If we identify p_t as the total energy of the particle divided by c^2, then the first term is the famous relation, equation [1.1], between energy and mass which started our discussion in Chapter 1. We will accept this hypothesis and in future $\vec{\mathbf{p}}$ will be referred to as the 'energy-momentum four-vector'.

The first property of this four-vector that we shall investigate is the value of its invariant, i.e. the value of its 'length'. Thus the square of the length is given by

$$\vec{p}^2 = (cp_t)^2 - p_x^2 - p_y^2 - p_z^2 = (cp_t)^2 - p^2 \qquad [5.11]$$

Substituting from equations [5.9], we find for the square of the length:

$$\vec{p}^2 = \frac{m_0^2}{1-(u/c)^2} (c^2 - u_x^2 - u_y^2 - u_z^2) = m_0^2 c^2 \qquad [5.12]$$

Thus the length of the energy-momentum four-vector is equal to the rest energy squared divided by c^2. If we put this result in equation [5.11], remembering that $p_t = E/c^2$ and rearranging accordingly, we obtain the relation between energy and momentum

$$E^2 = m_0^2 c^4 + c^2 p^2 \qquad [5.13]$$

This result is analogous to the Newtonian result

$$E = p^2/2m$$

which may be recovered from equation [5.13] by taking the square root of both sides and expanding in powers of (p/m_0c). The first term in the expansion gives the rest energy

$$E = m_0 c^2$$

whilst the second term is the Newtonian kinetic energy. This agrees with the conventional result, provided we agree that the energy is to be measured from the constant rest energy.

There is one particle in nature that has zero rest mass, the photon, and for it equation [5.12] tells us that the length of the energy-momentum four-vector is zero! (It is important not to think in terms of Euclidean geometry.) Further, the energy-momentum relation is particularly simple:

$$E = cp \qquad [5.14]$$

We jump ahread and postulate the Planck relation between energy and frequency

$$E = h\nu$$

and the de Broglie relation

$$p = h/\lambda$$

(We shall show in Chapter 6 that if we assume the validity of the principle of relativity, then it is only necessary to assume one of these relations; the other follows by deduction.) It now follows from equation [5.14] that

$$\nu\lambda = c$$

the well-known relation between frequency and wavelength for electromagnetic waves.

All this is very interesting but, apart from the result $E = m_0c^2$, we learn nothing unless we postulate a law for the energy-momentum four-vector. Suppose we have a collection of particles, each with its own energy-momentum four-vector $\vec{\mathbf{p}}_1, \vec{\mathbf{p}}_2, \vec{\mathbf{p}}_3 \ldots$. Then we follow the obvious course and postulate that the sum of the four vectors is a constant, i.e.

$$\vec{\mathbf{p}}_1 + \vec{\mathbf{p}}_2 + \vec{\mathbf{p}}_3 + \ldots = \vec{\mathbf{C}} \qquad [5.15]$$

Since the sum of four-vectors is also a four-vector, it follows from our analysis that led to equation [5.6] that such a law will be form invariant under the Lorentz transformation. However, the only real test of the validity of equation [5.15] is experiment; if it predicts some of the behaviour of, say, two interacting particles, then we have grounds for accepting its validity. On the other hand if it contradicts experiment, we shall simply have to try again!

Before we examine the experimental evidence for the validity of equation [5.15], it is interesting to examine the implications of the Lorentz transformation on the energy-momentum four-vector. If we put

$$p_t = E/c^2$$

in the transformation in equation [5.1], putting $F_x = p_x$, $F_y = p_y$, etc., we get

$$p_x' = \gamma(p_x - vE/c^2)$$
$$p_y' = p_y \qquad p_z' = p_z \qquad [5.16]$$
$$E' = \gamma(-vp_x + E)$$

This result again shows that we live in a four-dimensional world and that there is nothing special about any one of the dimensions. For this transformation we see that, for example, what is called energy in frame K' is an admixture of what is called energy and momentum in frame K. Similarly if we write down the inverse transformation (by the usual method of changing the sign of v and interchanging the primed and unprimed quantities) we get

$$\begin{aligned} p_x &= \gamma(p_x' + vE'/c^2) \\ p_y &= p_y' \qquad p_z = p_z' \\ E &= \gamma(vp_x' + E') \end{aligned} \qquad [5.17]$$

Now the opposite is true; what is now, say, energy in the frame K' is an admixture of what was energy and momentum in the frame K. Thus energy and momentum are inextricably mixed and, more importantly, the postulated law in equation [5.15] implies not only conservation of momentum but also, because of the time-like component, conservation of energy as well. This is a very interesting development; we find that in Einstein's special relativity we need less laws of physics. The reason for this is simply the properties of the Lorentz transformation; it takes us into a four-dimensional world and if the law holds in one dimension then it must hold in all four. This is the same statement as saying that the four-dimensional space is isotropic, i.e. it has the same properties in all directions. In Newtonian mechanics it is always assumed that conservation of momentum holds in all directions, i.e. that Euclidean space is isotropic.

We must now turn to investigate the experimental evidence for the conservation of the energy-momentum four-vector. This will also imply a test for the expressions for momentum and energy (equations [5.9]).

5.5 Analysis of experiments

Before we can analyse the experiments and compare their results with the theoretical expressions, we should first consider precisely the nature and type of experiment we wish to perform. To clarify the situation, let us go back to the Newtonian situation. In that regime, if we wish to find the paths in space travelled by a set of particles, then we have to solve a set of second-order differential equations. Equations [2.1] are an example of the dynamic equations for two particles interacting through a Hooke's law force. On the other hand if we do not require such detailed information, then we can get a partial solution by simply considering the equations of conservation of momentum and energy. A very familiar example of this approach is the case of the collision of two elastic spheres. Since both momentum and energy are conserved, the total momentum before the collision is equal to the total momentum after the collision, with a similar statement for energy. To introduce the notation we shall use in the general case, let $\mathbf{p}_1$ and $\mathbf{p}_2$ be the momenta of the two particles before the collision and $\mathbf{q}_1$, $\mathbf{q}_2$ the momenta after the collision. Then

$$\mathbf{p}_1 + \mathbf{p}_2 = \mathbf{q}_1 + \mathbf{q}_2 \qquad [5.18]$$

Although we are in the Newtonian regime, we will introduce the notation $p_t = E/c^2$ so that we can take our results over to the more general case. Thus conservation of energy gives

$$p_{t1} + p_{t2} = q_{t1} + q_{t2} \qquad [5.19]$$

Thus we have four equations (equation [5.18] has three components) with sixteen variables to be found, i.e. three components of momentum and the energy for each particle before the collision and a similar number after. Eight of the variables – the ones before the collision – are specified by the initial experimental conditions. We should note that, provided that the force between the particles is of sufficiently short range, it is possible to speak of before and after collision situations and the energies of the particles are simply their kinetic energies, i.e.

$$E = p^2/2m$$

Thus the number of unknowns is reduced to six, to be determined by four equations. Therefore the sort of questions we are able to answer are those where experiment has, say, determined two components of the momentum of one particle after the collision and we are asked to predict the momentum of the remaining particle. For example, suppose for two particles of equal mass the initial experimental set up is such that

$$\mathbf{p}_1 = \hat{\mathbf{i}} p_0$$

and $\mathbf{p}_2 = 0$

and that after the collision

$$q_{y1} = q_{z1} = 0$$

then the equations of conservation of momentum and energy become

$$p_0 = q_{x1} + q_{x2} \qquad 0 = q_{y2} \qquad 0 = q_{z2}$$

$$p_0^2 = q_{x1}^2 + q_{x2}^2$$

These equations have the solution

$$q_{x1} = q_{y2} = 0 \qquad q_{x2} = p_0.$$

(There is also a solution

$$\mathbf{q}_2 = 0 \qquad q_{x1} = p_0$$

but this implies that particle one has passed straight through particle two, i.e. there is no force of interaction between them.)

The situation is exactly the same when we move from the Newtonian regime to the general case, the only difference being the relationship between energy and momentum. Instead of the simple Newtonian relationship $E = p^2/2m$, we have the more complicated relation of equation [5.13]. However because the energy and momentum form a four-vector, the conservation equations can be written down as one four-vector equation. Further, it is, as we shall see, easy to develop techniques for solving the equations.

Before doing so we should look at the validity of the expressions [5.9] for the energy and momentum as functions of the particle velocity. To check the expressions for the momentum we would have to have the concept of force, which in the general case we do not have; in fact we shall postpone introducing force until Chapter 7. Therefore we shall just look at the expression for the energy. To check the validity of this expression we shall make an additional assumption that for one particle the sum of the kinetic and potential energies is a constant.

5.6 The Lorentz invariant expression for the energy

Consider an electron that has been accelerated through a voltage of V volts. With the assumption that all of the potential energy has been converted to kinetic energy, the Newtonian expression for its final velocity will be

$$u = (2eV/m)^{1/2} \qquad [5.20]$$

In the general Lorentz invariant case, equation [5.9] and the identity

$$p_t = E/c^2 \qquad E = m_0c^2 + eV$$

give for the final velocity

$$u/c = \left[1 - \left(\frac{m_0c^2}{eV + m_0c^2}\right)^2\right]^{1/2} \qquad [5.21]$$

If the kinetic energy of the particle (= eV) is large compared to the rest energy m_0c^2, then these two results are sufficiently different for the two velocities to be distinguished experimentally. For example, the synchrotron radiation source at the Daresbury nuclear physics laboratory in the north of England accelerates electrons to kinetic energies of the order GeV (1 GeV = 10^9 electron volts = 1.602×10^{-10} J). For 1 GeV, equation [5.20] gives a velocity of (the mass of the electron is 9.110×10^{-31} kg) 1.9×10^{10} m s^{-1}, i.e. nearly a hundred times faster than the velocity of light! On the other hand equation [5.21] gives

$$u/c = (1 - 2.5 \times 10^{-7})^{1/2} \simeq 1 - 2.5 \times 10^{-7}/2$$

or $$(c - u)/c = 1.25 \times 10^{-7}$$

a result which is very close to the velocity of light but still less than it. Results at Daresbury and other accelerator laboratories are consistent with the Lorentz invariant form of the energy rather than the Newtonian form. It is in fact true to say that, to date, no particle, whatever energy it has been accelerated to, has been found to have a velocity greater than the velocity of light.

With this encouraging result we now return to the general methods of analysis of collision experiments.

5.7 Techniques of analysis in two-particle collisions

The sort of experiments that we will be concerned with analysing can be illustrated symbolically by the reaction,

$$A + B \rightarrow C + D + \ldots$$

This is a generalization of the two-body collision considered in the previous section; it allows the possibility of two particles A and B colliding and producing a group of particles C, D, E, etc. This latter group of particles may contain none, one, or both of the original pair of colliding particles. Conservation of the energy-momentum four-vector can then be written

$$\vec{p}_A + \vec{p}_B = \vec{q}_C + \vec{q}_D + \ldots$$

where we continue to use the notation that four-vectors for particles after the collision are denoted by $\vec{q}$. The problem, as we discussed in section 5.5, is that we know the four-vectors for A and B and we are given some information about the four-vectors of C, D, etc., and we have to find the unknown momenta and energies. The solution to such a problem can be solved by the use of a particular technique.

Suppose that we take the case where only two particles are produced on the right-hand side, say C and D, which may or may not be identical to A and B. Further, suppose we are told nothing about the energy-momentum four-vector of particle D but we are asked a question about the dynamic state of particle C. The technique is to rearrange the equation for the conservation of the energy-momentum four-vector so that the four-vector for the particle which we are not interested in, i.e. particle D, stands on its own on one side of the equation. Thus

$$\vec{q}_D = \vec{p}_A + \vec{p}_B - \vec{q}_C$$

Square both sides (strictly speaking take the scalar product of each side with itself) and use the result that the 'length' squared of the energy-momentum four-vector is an invariant and equal to $(m_D c)^2$ (see equation [5.12]) to obtain

$$(m_D c)^2 = (m_A c)^2 + (m_B c)^2 + (m_C c)^2 + \\ + 2\vec{p}_A \cdot \vec{p}_B - 2\vec{p}_A \cdot \vec{q}_C - 2\vec{p}_B \cdot \vec{q}_C$$

From this expression it will be seen that the only four-vectors remaining are those that we know, i.e. $\vec{p}_A$ and $\vec{p}_B$, or the one we wish to find, $\vec{q}_C$. This is as far as we can sensibly take the general case, and the most instructive thing to do next is to look at a specific example.

5.8 Spontaneous decay of an excited atom

Consider the case of an excited atom, denoted by A*, which decays spontaneously to the ground-state atom A with the emission of a photon γ, i.e.

$$A^* \rightarrow A + \gamma$$

We choose the x direction as the direction of the momentum of the excited atom and ask the question, how does the energy of the emitted photon depend upon the angle of emission? The situation is shown in Fig. 5.1. Conservation of the energy-momentum four-vector gives

$$\vec{p}_{A^*} = \vec{q}_A + \vec{q}_\gamma$$

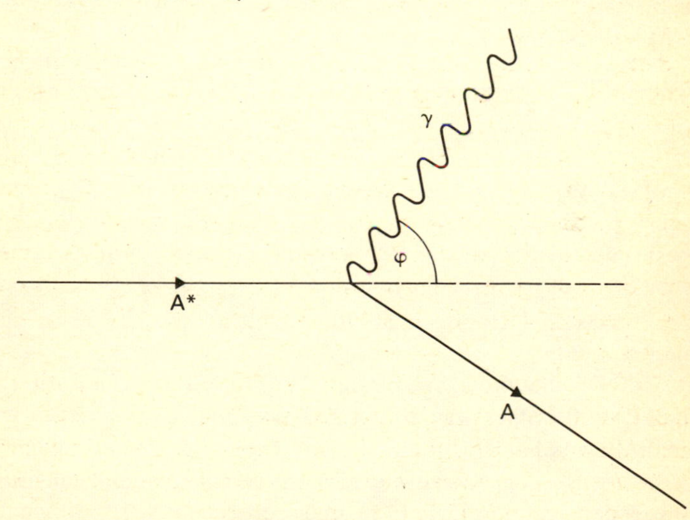

Fig. 5.1 An excited atom A* emits a photon γ and returns to its ground state A.

We are not interested in the dynamic state of the ground-state atom, so we rearrange this equation so that the four-vector for that particle is on the left-hand side

$$\vec{q}_A = \vec{p}_{A*} - \vec{q}_\gamma$$

Taking the 'square' of both sides gives us

$$\vec{p}_A \cdot \vec{p}_A = m_A^2 c^2 = m_A^{*2} c^2 + 2\,\mathbf{p}_{A*} \cdot \mathbf{q}_\gamma$$

since the mass of the photon is zero. The scalar product of two four-vectors is defined by equation [5.3]; thus

$$m_A^2 c^2 = m_{A*}^2 c^2 - 2c^2 (E_{A*}/c^2)(E_\gamma/c^2) + 2\mathbf{p}_{A*} \cdot \mathbf{q}_\gamma$$

Now for a photon equation [5.14] holds, again because the rest mass is zero, i.e.

$$q_\gamma = E_\gamma/c$$

and if ϕ is the angle between the directions of the emitted photon and the incoming excited atom then

$$m_A^2 c^2 = m_{A*}^2 c^2 - 2c^2 (E_{A*}/c^2)(E_\gamma/c^2) + 2p_{A*} (E_\gamma/c) \cos\phi$$

which on rearrangement becomes

$$E_\gamma = \frac{(m_{A*}^2 - m_A^2)c^4}{2(E_{A*} - cp_{A*} \cos\phi)} \qquad [5.22]$$

This simple calculation shows some interesting and possibly unexpected results. First, the excited atom must have a larger rest mass than the ground-state atom, for, if we were to put $m_{A*} = m_A$ then the energy of the photon would be zero. However, we know that an excited atom can and does emit a photon of finite and non-zero energy.

To see the order of magnitude involved we can turn the question the other way round and ask, for a photon of a given frequency, what is the mass difference? As an example, the so-called H_α line of the emission spectrum of hydrogen has a wavelength of 6563×10^{-10} m and consequently, according to quantum mechanics, $E = h\nu = hc/\lambda = 3.0 \times 10^{-19}$ J $= 1.9$ eV. Given that the mass of the hydrogen atom is 939 MeV, this means that $(m_{A*} - m_A)/m_A \sim 10^{-8}$, certainly beyond the reach of any direct determination of mass.

Second, it is possible to obtain a simple interpretation of the result in equation [5.22]. Suppose, initially, that the excited atom is at rest; then according to equation [5.13]

$$E_{A*} = m_{A*}c^2$$

and thus equation [5.22] simplifies to

$$E^0_\gamma = (m^2_{A*} - m^2_A)c^2/2m_{A*}$$

Thus the frequency of the emitted photon, when the excited atom is at rest, is

$$\nu_0 = E^0_\gamma/h = (m^2_{A*} - m^2_A)c^2/2m_{A*}h \qquad [5.23]$$

When the excited atom is in motion its energy and momentum are related to its velocity by equations [5.9]. If u, the excited atom's velocity, is in the x direction, then the results [5.23] and [5.22] give for the frequency of the emitted photon when the excited atom is moving

$$\nu_0 \frac{[1-(u/c)^2]^{1/2}}{[1-(u \cos\phi)/c]} \qquad [5.24]$$

If we assume that u/c is small, as it usually is, then equation [5.24] can be expanded to give

$$\nu \simeq \nu_0[1 + (u\cos\phi)/c] \qquad [5.25]$$

This formula is readily recognizable as the frequency of a wave when the source is moving with a velocity u and emitting radiation at an angle ϕ to its direction of travel, given that the frequency was ν_0 when the source was at rest, i.e. the frequency is Doppler shifted. This effect is known to every spectroscopist; the atoms of a gas are all moving randomly with different velocities and consequently the frequencies emitted from different excited atoms will be different (even though the atoms are excited to the same state). Thus instead of observing a single line, the line will be broadened into a band. This effect is known as Doppler broadening. To find the order of magnitude, we take the kinetic energy of a single atom to be of the order of $3\,kT/2$. For temperatures of the order of 300 K, and again taking the hydrogen atom, we find the velocity $u = (3\,kT/m)^{1/2} \sim 2.8 \times 10^3$ m s^{-1} (at these sorts of velocities it is permissible to use the Newtonian expression for the kinetic energy). Hence $u/c \sim 10^{-5}$ and thus $(\nu - \nu_0)/\nu_0 \sim 10^{-5}$, where we have taken the case $\cos\phi = 1$. Such differences are easily detectable by modern spectroscopic techniques.

We should note in passing that equation [5.24] is the exact expression for the shifted frequency. We shall show in Chapter 6 that this is in fact the correct expression for the Doppler shift when time dilation is taken into account. In the meantime we consider a second example of the conservation of the energy-momentum four-vector.

5.9 Positron-electron annihilation

Under certain circumstances a positron (a particle of the same mass as the electron but of opposite charge) and an electron will annihilate to produce two photons. The reaction is illustrated in Fig. 5.2 for the special case where the positron is at rest. The question we will ask is what is the energy of photon 1, as a function of the angle between the two photons? Conservation of the energy-momentum four-vector gives

$$\vec{p}_+ + \vec{p}_- = \vec{q}_{\gamma_1} + \vec{q}_{\gamma_2} \qquad [5.26]$$

We are not interested in photon 2 and therefore take all other four-vectors to one side and then take the 'square' of both sides:

$$\begin{aligned} \vec{q}^{\,2}_{\gamma_2} &= \vec{q}_{\gamma_2} \cdot \vec{q}_{\gamma_2} \\ &= \vec{p}_+ \cdot \vec{p}_+ + \vec{p}_- \cdot \vec{p}_- + \vec{q}_{\gamma_1} \cdot \vec{q}_{\gamma_1} + 2\vec{p}_+ \cdot \vec{p}_- \\ &\quad - 2\vec{p}_+ \cdot \vec{q}_{\gamma_1} - 2\vec{p}_- \cdot \vec{q}_{\gamma 1} \end{aligned} \qquad [5.27]$$

To evaluate this expression we need the following results that have already been given:

1. the definition of the scalar product (equation [5.3]);
2. the zero mass of the photon;
3. the relation between energy and momentum for a photon (equation [5.14]); and
4. the invariance property of the energy-momentum four-vector for a single particle (equation [5.12]).

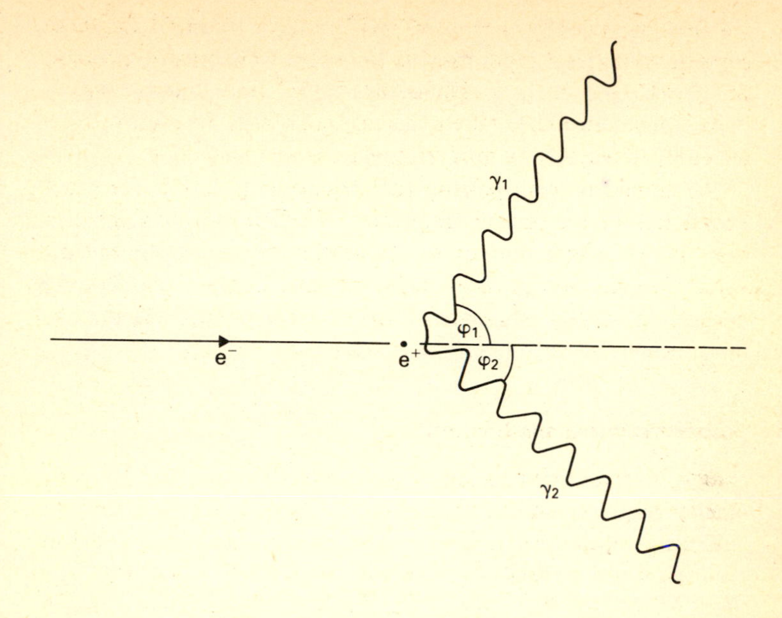

Fig. 5.2 An electron e⁻ with a constant velocity interacts with a stationary positron e⁺, to produce two photons γ_1 and γ_2. This process is known as electron–positron annihilation.

Then, since the positron is initially at rest and consequently $\mathbf{p}_+ = 0$ and $E_+ = mc^2$, equation [5.27] reduces to

$$\begin{aligned} 0 = {} & m^2c^2 + m^2c^2 + 2c^2(E_-/c^2)(mc^2/c^2) \\ & - 2c^2(E_+/c^2)(E_{\gamma_1}/c^2) - 2c^2(E_-/c^2)(E_{\gamma_1}/c^2) \\ & + 2\mathbf{p}_- \cdot \mathbf{q}_{\gamma_1} \end{aligned}$$

If ϕ_1 is the angle between the direction of propagation of photon 1 and the direction of the incoming electron, so that

$$\mathbf{p}_- \cdot \mathbf{q}_{\gamma_1} = p_- q_{\gamma_1} \cos\phi_1 = p_- E_{\gamma_1} \cos\phi/c$$

then a rearrangement gives

$$E_{\gamma_1} = \frac{mc^2(E_- + mc^2)}{(E_- + mc^2) - cp_- \cos\phi_1} \qquad [5.28]$$

For fixed values of E_- and p_- the photon energy will be at a maximum when $\cos\phi_1 = 1$, i.e. the photon is emitted in the forward direction and is at a minimum when it is emitted in the backward direction, i.e. $\cos\phi_1 = -1$.

Finally we should note the obvious symmetry in the problem. There is nothing special about photon 1; our results apply equally well to photon 2. Thus if photon 1 is emitted in the forward direction, then, because of conservation of energy and momentum and because the initial momentum, by definition, is in the forward direction, it follows that photon 2 must be emitted in the backward direction with the minimum energy.

5.10 Compton scattering

As a final example of two-body collisions we consider the so-called Compton scattering, where an electron at rest scatters an incident photon (Fig. 5.3 shows the general case where the initial electron is moving, see exercise 4). The problem is, given

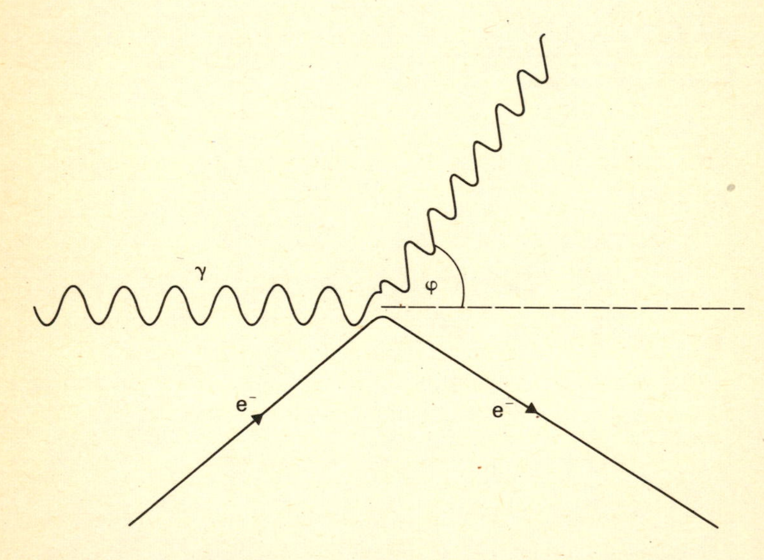

Fig. 5.3 A photon γ Compton scatters off an electron e^-. After the scattering, the photon has a different wavelength.

the incident photon energy, what is the energy of the scattered photon? In this experiment we are not interested in the electron after the scattering. Hence the energy-momentum four-vector conservation equation is written

$$\vec{q}_- = \vec{p}_- + \vec{p}_{\gamma 1} - \vec{q}_{\gamma 2} \qquad [5.29]$$

Following the usual procedure, we take the scalar product of each side with itself, use the definition of the scalar product, the invariance of the square of the length and the special properties of the photon four-vector to obtain

$$m^2c^2 = m^2c^2 + 2c^2(E_{\gamma_1}/c^2)(mc^2/c^2) - 2c^2(E_{\gamma_1}/c^2)(E_{\gamma_2}/c^2) + 2\mathbf{p}_{\gamma_1}\cdot\mathbf{q}_{\gamma_2} - 2c^2(mc^2/c^2)(E_{\gamma_2}/c^2)$$

where E_{γ_1} and E_{γ_2} are the energies of the photon before and after the scattering respectively. Thus if ϕ is the angle between the scattered photon and the incident photon, then

$$\mathbf{p}_{\gamma_1}\cdot\mathbf{q}_{\gamma_2} = p_{\gamma_1}q_{\gamma_2}\cos\phi = (E_{\gamma_1}E_{\gamma_2}/c^2)\cos\phi$$

Hence if we solve for E_{γ_2} we find

$$E_{\gamma_2} = \frac{E_{\gamma_1}}{1 + (E_{\gamma_1}/mc^2)(1 - \cos\phi)} \qquad [5.30]$$

Now for a photon $E = hc/\lambda$ and consequently equation [5.30] can be written

$$\lambda_2 = \lambda_1[1 + (h/mc\lambda_1)(1 - \cos\phi)] \qquad [5.31]$$

or alternatively

$$\lambda_2 - \lambda_1 = \lambda_e(1 - \cos\phi) \qquad [5.32]$$

where

$$\lambda_e = h/mc$$

is known as the Compton wavelength.

The Compton effect is frequently quoted as an example of the experimental evidence for the breakdown of classical mechanics and hence the need to introduce quantum mechanics. In classical mechanics the photon would be treated as a wave and not as a particle, and consequently there would be no change in the wavelength after scattering.

Other examples of this type of application of the conservation of the energy-momentum four-vector are given in the exercises for this chapter (see Appendix 3). We now turn our attention to a second application of the law.

5.11 Collision threshold energies

The reader will have gathered by now that the majority of experiments involving elementary particles require one particle to strike another with the subsequent production of one, two or more particles of the same or different types as the initial ones. If, as the result of the interaction, the total mass of the particles produced is greater than the mass of the particles producing the interaction, then it is clear that some of the incoming kinetic energy must be converted to mass. If there is not sufficient incoming kinetic energy to provide the necessary mass deficit, then the reaction will not be able to occur. This reasoning leads to the concept of the threshold energy for a reaction, i.e. the minimum kinetic energy of the incoming particle so that the reaction will just occur.

At first glance it might seem easy to write down a criterion for the threshold energy. Consider a particular reaction

$$\mathrm{p} + \mathrm{p} \rightarrow \mathrm{p} + \mathrm{n} + \pi^+$$

where p stands for a proton, n for a neutron and π^+ for a positively charged pion. Suppose that the target proton is at rest. The question is then: What is the minimum energy, E_0, of the incoming proton, so that is has sufficient kinetic energy, equal to $E_0 - m_\mathrm{p}c^2$, for the proton, the neutron and the pion to be produced? The masses involved are m_p = 938 MeV $\simeq m_\mathrm{n}$ and m_π = 139 MeV, so that the sum of the masses after the interaction is 139 MeV greater than the sum of the masses before the interaction. For the threshold energy we need to find the energy when the particles are just produced and there is no excess energy for the kinetic energy of the reaction products.

It would seem, therefore, that for each of the particles produced by the reaction its energy-momentum four-vector should have the form where the time-like component should be the rest mass (the rest energy divided by c^2) and the space-like components, i.e. the three-momentum, should be identically zero. However, this is incorrect; the incoming proton has a non-zero momentum and therefore the total momentum after the reaction cannot be zero. We could have used the above specification of the four-vectors if the two initial protons were coming towards each other with equal and opposite momenta. Then the total momentum after the interaction would have to be zero and one possibility would be that the momenta of each of the reaction products would be identically zero. This situation would be achieved if we make a transition from a description of the experiment in the laboratory frame to a description in the so-called zero momentum frame (sometimes misleadingly referred to as the centre of mass frame). That is we consider a frame which is moving with a constant velocity with respect to the laboratory and the velocity is chosen so that the initial two protons are moving towards each other, as observed from the new frame, with equal and opposite momenta.

That this can always be done, irrespective of whether the initial colliding particles have the same mass, as in the above example, or not, is easily proved. Let the total momentum and total energy, as measured in the laboratory frame, of the initial colliding particles be **P** and E respectively. In the moving frame let the corresponding quantities be $\mathbf{P}'$ and E'. Then according to the Lorentz transformation

$$P_x' = \gamma(P_{x'} - vE/c)$$

If we take $P_{y'} = P_{z'} = 0$, then in the new frame all the momentum is also in the x' direction. The requirement for the zero-momentum frame is therefore that $P_x' = 0$. It follows therefore that v must be chosen so that

$$v = Pc/E \qquad [5.33]$$

From equation [5.13] it follows that $cP < E$ and therefore that $v < c$. Thus it is always possible to make the transition to the zero-momentum frame. In this frame the threshold energy is found by specifying that the momentum of each particle after the reaction is zero.

To see how this leads to the calculation of the threshold energy in the laboratory frame, we return to the specific example that started this discussion. At threshold in the zero-momentum frame, the total energy-momentum four-vector after the reaction is

$$\vec{\mathbf{Q}} = (0, m_\mathrm{p} + m_\mathrm{n} + m_\pi) \qquad [5.34]$$

The scalar product of this vector with itself is easily obtained and is of course just c^2 times the square of the time-like component. The argument now proceeds in two stages. First we use the conservation of the total four-vector to equate the above four-vector to the total four-vector before the collision as measured in the zero-momentum frame. Since the four-vectors are equal, it follows that the squares of their 'lengths' are also equal. The second stage of the argument is to use the fact that the square of the 'length' of a four-vector is an invariant under a Lorentz transformation. Thus the 'length' of the energy-momentum four-vector before the reaction in the zero-momentum frame must be equal to the 'length' of the energy-momentum four-vector in the laboratory frame. Thus stage one of the argument gives

$$\vec{P}'^2 = \vec{Q}'^2$$

i.e. conservation of the four-vector in the zero-momentum frame, whilst the second stage gives

$$\vec{P}^2 = \vec{P}'^2$$

i.e. the invariant property of the four-vector. Combining these two results gives

$$\vec{P}^2 = \vec{Q}'^2 \qquad [5.35]$$

$\vec{\mathbf{P}}$, the total energy-momentum four-vector before the collision, has a time-like component of $(E_0/c^2) + m_p$ and a space-like component solely in the x direction and equal to p. The square of its 'length' is therefore

$$\vec{P}^2 = c^2(m_\mathrm{p} + E_0/c^2)^2 - p^2$$

which, using the relation between energy and momentum for a particle, i.e.

$$E_0^{\,2} = m_{\mathrm{p}}^2c^4 + c^2p^2$$

reduces to

$$\vec{P}^2 = 2m_{\mathrm{p}}(E_0 + m_{\mathrm{p}}c^2) \qquad [5.36]$$

Thus combining equations [5.34], [5.35] and [5.36] we get

$$2m_{\mathrm{p}}(E_0 + m_{\mathrm{p}}c^2) = (m_{\mathrm{p}} + m_{\mathrm{n}} + m_\pi)^2 c$$

or, solving for E_0,

$$E_0 = \frac{(m_{\mathrm{p}} + m_{\mathrm{n}} + m_\pi)^2c^2}{2m_{\mathrm{p}}} - m_{\mathrm{p}}c^2 \qquad [5.37]$$

For the masses given above, this formula gives a value for E of 1226 MeV. The threshold energy is this value minus the rest energy of the proton, i.e. $1226 - 938 = 288$ MeV.

The above procedure may seem complicated on first reading, but this is not really the case. The reader is urged to do the exercises for this chapter (see Appendix 3), using the above example for guidance.

This then concludes our examples of the use of the postulated expression for the four-momentum and its conservation law. Space does not permit a description of the experiments that have been performed to verify the results given here, and similar calculations on other reactions. Suffice it to say that, to date, whatever discrepancies may have been found between theory and experiment, on no occasion has it been felt necessary to abandon the expression for the energy-momentum four-vector or its conservation law. As we remarked earlier in this chapter, the conservation of energy and momentum will not give a completely detailed description of a dynamic problem; for that task we need the Lorentz invariant form of the equivalent of Newton's equations of motion. The formulation of these equations is postponed to Chapter 7. In the next chapter we digress to look at the effect of the Lorentz transformation on the phenomena of waves.

Chapter 6
Wave propagation

6.1 Introduction

In the last chapter we formulated relativistically invariant forms for the law of conservation of momentum and energy. To follow a logical ordering we ought now to try to formulate the equivalent of the Newtonian equations of motion. As we indicated, however, this is by no means an easy task. The concept of action at a distance has to be abandoned and if one particle is to influence another, the information has to be conveyed from one to the other with a velocity whose maximum value is the velocity of light. The only forces we have considered in detail up until now have been electromagnetic ones. We have seen that the fields that are responsible for these forces are propagated through space as electromagnetic waves. This suggests that, at least as far as the electromagnetic forces are concerned, 'action at a distance' may be accomplished through the medium of waves propagating with the velocity of light. Therefore before we attempt to look for further dynamic laws, we will examine the propagation of waves as viewed from different inertial frames. We start by examining the propagation of waves in free space.

6.2 Waves in free space

We can derive some very interesting results about the propagation of waves in free space without, as yet, specifying the nature of those waves. We know that electromagnetic waves propagate in free space but it turns out that in quantum mechanics particles propagate as waves. We have, therefore, at least two examples of wave propagation in free space. In general a wave has an amplitude and a phase. Depending upon the type of wave the amplitude can be a vector or a scalar. (By 'vector' in this context we mean either a three-vector or a four-vector.) For a monochromatic wave, i.e. a wave of a well-defined angular frequency, the phase can be written

$$\phi = \omega t - \mathbf{k}\cdot\mathbf{r} \qquad [6.1]$$

Here ω is the angular frequency, $\mathbf{k}$ the wave vector, t the time and $\mathbf{r}$ the position vector. Thus in general we may write for the wave

$$\psi = C \exp i(\omega t - \mathbf{k}\cdot\mathbf{r}) \qquad [6.2]$$

where C is a scalar or a vector according to the nature of the wave.

How does the wave transform under the Lorentz transformation? To answer this question we have to know how the two quantities, the amplitude and the phase, transform. For the first of these we have to know the physics applicable to the particular wave we are considering; we have already seen how the electric and magnetic fields transform under the Galilean transformation and in Chapter 7 we shall show how they transform under the Lorentz transformation. In both of these cases we can only do this by considering the physical properties of the field. The phase is, however, a scalar, whatever the physical nature of the wave. A quite general result for the Lorentz transformation, and for that matter the Galilean transformation, is that a scalar has the same value in all inertial frames, i.e. it is an invariant. (This argument, although frequently quoted, is not rigorous. *A priori* there is no reason why ϕ, although a scalar, could not be one component of a four-vector, however unlikely this may seem. Nevertheless it is a possibility and, if it were so, ϕ would not have the same value in all inertial frames. For a more rigorous argument see C. Moller, *The Theory of Special Relativity*, Oxford University Press, 1957.) We can therefore investigate the property of the invariance of the phase irrespective of the physical nature of the wave.

Since the phase (equation [6.1]) depends upon x, y, z, t, the phase in the inertial frame K' will depend upon x', y', z', t', where the two sets of coordinates – the unprimed and the primed – are related by the Lorentz transformation. The invariance property is thus expressed as

$$\omega t - k_x x - k_y y - k_z z = \omega' t' - k'_x x' - k'_y y' - k'_z z' \quad [6.3]$$

Using the inverse Lorentz transformation (equation [4.7]) and collecting together the coefficients of x', y', etc., we find

$$k'_y = k_y \qquad k'_z = k_z$$

$$k'_x = \gamma(k_x - v\omega/c^2) \quad [6.4]$$

$$\omega'/c^2 = \gamma(-k_x v/c^2 + \omega/c^2)$$

Inspection of equation [6.3] shows that the phase in the new inertial frame, K', has the same form as the phase in the old inertial frame, K, but with a new frequency ω' and wave numbers k'_x, k'_y, k'_z. Equations [6.4] show that these quantities are related to the corresponding quantities in the original frame, K, by a Lorentz transformatin, i.e. ω/c^2 and **k** form a four-vector. In fact we could have anticipated this result. Since **r** and t form a four-vector, the form of the phase is that of the scalar product of two four-vectors (see equation [5.3]). Since this quantity has the same value in all inertial frames, we can deduce that **k** and ω/c^2 form a four-vector.

This result means that we can make our first prediction about the propagation of waves in free space. In Chapter 5 we showed that the square of the 'length' of a four-vector was a scalar invariant. Equation [5.2] gives the definition of that quantity and thus for the frequency wave vector four-vector we have

$$(\omega/c)^2 - k^2 = A \quad [6.5]$$

where A is a constant which is characteristic for the particular wave under consideration. A cannot be a pure number (unless it is identically zero) since it has to have the dimensions of length to the minus two. The question then remains, is it a universal constant or does it depend upon the particular type of wave under consideration? We cannot answer this question until we look at specific examples of waves propagating in free space.

Returning to equation [6.5] and rearranging, we find the dispersion relation for free space:

$$\omega = (A + k^2)^{1/2} c \quad [6.6]$$

and hence ω and k are not independent variables. We should also note that A can, in general, be either positive or negative, in the same way that the interval of the space-time coordinates could take either sign, depending upon whether the interval was space-like or time-like.

Thus the invariance of the phase under the Lorentz transformation tells us that any wave that propagates in free space, irrespective of its nature, must have a dispersion relation of the form in equation [6.6]. The dependence of the constant A on any parameters will depend on the particular waves under consideration. For example, if we are to obtain the correct dispersion relation

$$\omega = ck$$

for electromagnetic waves, then we know that the value of A must be zero.

A correct theory should, of course, be able to predict the value of the constant A rather than giving it a value to obtain the correct result. The invariance of the phase has yielded as much information as is possible and if we are to get any further then we have to introduce some more physics. Chapter 5 was concerned with a relativistically invariant mechanics and it would be attractive if we could exploit this in our theory of waves. Fortunately the other great triumph of late nineteenth- and early twentieth-century physics—quantum mechanics—enables us to do just that.

Quantum mechanics

It should have been noted by the reader that the result that frequency and wave vector form a four-vector is another example where, under a Galilean transformation, these quantities would have been distinct, but under the Lorentz transformation they are inextricably connected in the same way as space and time and momentum and energy. This suggests that we really need only one result from quantum mechanics. The simplest one to take is the Planck postulate that energy and frequency are related by

$$E = \hbar\omega \qquad [6.7]$$

Further, we now invert our usual argument and, rather than insist that this law must satisfy the principle of relativity, we assume that it does and look at the consequences. The last of equations [5.16] gives us the value of the energy in the frame K':

$$E'/c^2 = \gamma(-vp_x/c^2 + E/c^2)$$

whilst the last of equations [6.4] gives us the corresponding frequency in the same frame:

$$\omega'/c^2 = \gamma(-vk_x/c^2 + \omega/c^2)$$

Because we are assuming that the Planck relation holds in all inertial frames, it follows that

$$-vp_x + E = -\hbar v k_x + \hbar\omega$$

and, since $E = \hbar\omega$, we have

$$p_x = \hbar k_x \qquad [6.8]$$

which is the de Broglie relation. Before going on to discuss this result, we should remember that this is a prediction, unlike the Planck relation, which was a postulate. This means that if we are to accept the validity of the de Broglie relation, we must ensure that it is form invariant under the Lorentz transformation, a property that we assumed for the Planck relation. To show the form invariance we need the inverse of the transformation given by equations [6.4]. This we get in the usual way by interchanging the primed and unprimed quantities and changing the sign of v:

$$\begin{aligned} k_y &= k'_y \qquad k_z = k'_z \\ k_x &= \gamma(k'_x + v\omega/c^2) \qquad [6.9] \\ \omega/c^2 &= \gamma(vk'_x/c^2 + \omega'/c^2) \end{aligned}$$

In particular we need the last of these equations and the last of equations [5.17]. Identical arguments to those given above show

$$p'_x = \hbar k'_x$$

Hence the relation is form invariant in all inertial frames. Thus the assumption of the Planck relation is, because of the mixing of both momentum and energy and wave vector and frequency, sufficient to establish the validity of the de Broglie relation. If we had found that the de Broglie relation was not form invariant then we would have had to abandon the Planck relation and the assumption that it was form invariant in all inertial frames. We have, of course, only proved the result for the x component of the wave vector $\mathbf{k}$. However, if we had performed the Lorentz transformation along the y, rather than the x axis, then we would have shown that

$$p_y = \hbar k_y$$

and similarly for the z direction.

These two relations, Planck's and de Broglie's, show that the particle and the wave are two manifestations of the same entity. This is precisely the connection that we needed to be able to use the results of Chapter 5 in our theory of waves in free space. In particular, if we use the de Broglie and Planck relations in the energy-momentum relationship for a particle (equation [5.13]) and combine it with equation [6.5], we obtain

$$A = (m_0 c/\hbar)^2 \qquad [6.10]$$

which is the Planck frequency of the rest energy divided by c^2, all squared. So the simple quantum mechanical law $E = \hbar\omega$ has enabled us to derive the only unknown in the dispersion relation (equation [6.6]). Hence the dispersion relation is

$$\omega = c[(m_0 c/\hbar)^2 + k^2]^{1/2} \qquad [6.11]$$

The result in equation [6.10] is a very interesting one; it tells us how the constant A depends upon the parameter of the particular waves under consideration, i.e. it depends upon the rest mass of the particle. A little thought would have shown that the rest mass was the only parameter that could have entered into the expression for A. There are no other parameters that describe the particle waves which are independent of the motion of the particle. In other words we would have expected A to depend upon parameters in the rest frame of the particle. In another frame, any other (than the rest mass) parameters would depend upon the particle motion and would not therefore be constant. We labour this point because it will be a useful guide when we come to examine waves propagating in a medium.

From now on we assume that the wave is propagating in the x direction, so that

$$k_y = k_z = 0 \qquad k_x = k$$

The phase velocity

$$v_{ph} = \omega/k$$

and the group velocity

$$v_{gr} = d\omega/dk$$

can now be written down:

$$v_{ph} = c[(m_0c/\hbar)^2 + k^2]^{1/2}/k \qquad [6.12]$$

$$v_{gr} = ck/[(m_0c/\hbar)^2 + k^2]^{1/2} \qquad [6.13]$$

The first of these velocities is greater than c but the second is less than c. This latter result is important because it is the group velocity with which information is conveyed. A plane sinusoidal wave cannot convey information; if we wish to convey information we have to modulate the wave. The modulation of a wave travels with the group velocity.

These are fascinating results; they tell us that for a particle of zero rest mass i.e., a photon, the phase velocity is equal to the group velocity and both have a magnitude equal to the velocity of light. In all other cases the phase velocity is greater than the velocity of light. This latter result is consistent with our earlier result that information cannot be conveyed at a velocity greater than the velocity of light. This result can be amplified by looking at the relation between the group velocity and the classical particle velocity.

If we use the de Broglie relation in equation [6.13], we find for the group velocity

$$v_{gr} = c^2p/(m_0^2c^4 + c^2p^2)^{1/2}$$

But from equation [5.13] the denominator is just the total energy of the particle. If we therefore use equation [5.9], which relates the energy and momentum to the particle velocity, we find that

$$v_{gr} = u \qquad [6.14]$$

Thus the group velocity is identical to the 'particle' velocity. If we are conveying information by means of particles, then this is the result we would expect.

The only particle definitely known to have a zero rest mass is the photon. Until recently it was believed that the neutrino had zero rest mass. However astronomers now believe that there is not sufficient mass in the universe to be consistent with the current cosmological theories. Attempts to measure the mass of the neutrino have been made and the latest result at the time of writing is a rest mass of 35 ± 2 eV (see exercise 2, Chapter 5, Appendix 3). There is by no means universal agreement that this result is correct and confirmation or refutation will have to await further experiments. The photon is, of course, the particle aspect of electromagnetic radiation; consequently its phase velocity should be the velocity of light.

To complete this section we derive the Klein-Gordon equation, which is the wave equation satisfied by some particles. If we operate with the so-called d'Alembertian

$$\nabla^2 - \frac{1}{c^2}\frac{\partial^2}{\partial t^2} = \frac{\partial^2}{\partial x^2} + \frac{\partial^2}{\partial y^2} + \frac{\partial^2}{\partial z^2} - \frac{1}{c^2}\frac{\partial^2}{\partial t^2}$$

on the wave denoted by equation [6.2], then we obtain

$$[-k^2 + (\omega/c)^2]\,\psi$$

Thus using the dispersion relation (equation [6.11]) we find that ψ satisfies the wave equation

$$\nabla^2\psi - \frac{1}{c^2}\frac{\partial^2\psi}{\partial t^2} = \left(\frac{m_0 c}{\hbar}\right)^2 \psi \qquad [6.15]$$

This equation is known as the Klein-Gordon equation. It is left as an exercise for the reader to show that it is form invariant under a Lorentz transformation. We should remember that ψ is a scalar field. This latter assumption does not hold universally; some particles in nature have to be described by a vector field. Consideration of such matters would take us far beyond the scope and level of this book.

That, then, ends our discursion into quantum mechanics. We

have only had space to point out the direction, rather than explore the field, but hopefully the reader's interest will have been sufficiently aroused to explore matters further.

6.4 Waves propagating in a medium

So far in this chapter we have been concerned with the propagation of waves in free space, where all inertial frames are equivalent and the principle of relativity applies. Our analysis of this problem was based on the invariance of the phase from one inertial frame to another. We should note, however, that at no point in the argument did we invoke the principle of relativity. Now consider some physical wave which is propagating in a medium. It is perfectly possible to analyse the propagation of the wave from two different reference frames which are moving with a constant velocity with respect to each other. We would not expect the principle of relativity to apply, since the two frames, from the point of view of the propagation of the waves, are clearly distinguishable. In one frame the medium of propagation could, for example, be at rest whilst in the other it would be moving with the relative velocity of the two frames. In particular, in going from one frame to the other we would make use of the Lorentz transformation. The form of the wave would still be that given by equation [6.2], and in particular the argument given above for the invariance of the phase applies just as well in a medium as it does in free space. It follows, therefore, that much of the analysis given in section 6.3 applies to this situation; in fact the problem at first sight is to detect any differences between the two cases!

In considering waves in free space, the invariance of the phase led to the conclusion that the wave vector and the frequency (divided by c^2) formed the components of a four-vector. This result must also hold for waves propagating in a medium. So, in addition, must the dispersion relation (equation [6.6]), since this result follows from the invariance of the 'length' of the four-vector. The difference between the free-space case and the case of waves propagating in a medium must therefore lie in the constant A. Constant is perhaps a misnomer; if A really is a constant then all waves, whatever their nature, would obey the dispersion relation associated with the Klein-Gordon equation. There are many examples of dispersion relations in physics which are manifestly not of this form.

What this implies is that A is not strictly a constant but must depend upon some parameters. We can get some idea of which parameters by going back to the propagation of matter waves in free space. There A depended on the rest mass and we argued that this had to be because the rest frame of the particle was a special frame from the point of view of describing the particle. For photons there exists no rest frame and since A has dimensions of length to the minus two, it cannot be a pure number and therefore it must be zero. Now for waves propagating in a medium there is also a special frame, namely the frame in which the medium is at rest. We might therefore expect A to depend upon the frequency of the wave when the medium is at rest. If we denote, for this section only, the frequency and wave number in the rest frame by ω_0 and k_0 respectively, then from equation [6.6] the dispersion relation for waves propagating in a medium is

$$(\omega/c)^2 - k^2 = A(\omega_0) \qquad [6.16]$$

In particular in the rest frame of the medium

$$(\omega_0/c)^2 - k_0^2 = A(\omega_0) \qquad [6.17]$$

From equation [6.17] we can deduce the phase velocity in the rest frame of the medium:

$$v_{\mathrm{ph}_0} = \omega_0/k_0 = c/[1 - c^2 A(\omega_0)/\omega_0^2]^{1/2} \qquad [6.18]$$

This is as far as we can go without again having to introduce some more physics. In the case of free space we used the Planck and de Broglie relations, plus the expressions for the energy and momentum of a particle which we had derived so that they were consistent with the principle of relativity. Now that we have a medium present we cannot use the principle of relativity, and in that sense we should not have included a discussion of this phenomenon in this book. However, it is useful to recognize that we have used both the Lorentz transformation and the idea of a four-vector. Neither of these tools requires us to invoke the principle of relativity. This is perhaps an important distinction to make. The Lorentz transformation enables us to relate quantities in one frame to the same quantities in another frame that is related to the first by a constant relative velocity. The four-vector is a useful mathematical concept, the invariance of its length following from the properties of the Lorentz transformation. The principle of relativity is a physical postulate that says that the laws of physics should be form invariant in all inertial frames. To use the principle of relativity we may make use of the Lorentz transformation or the properties of four-vectors, even though they are different concepts, the first being a physical property, the second mathematical tools.

Nevertheless, despite the fact that we are strictly speaking outside our brief, it is interesting to complete the story of waves propagating in a medium. To do this we have to have a model for the particular medium we are considering and to decide what sort of waves are propagating. To be specific we shall give a very brief outline of the propagation of electromagnetic waves in a dielectric and in a metal. We consider first the case of the dielectric.

The phenomenological description of electromagnetic waves propagating in a medium is usually given in terms of the refractive index. Thus if $n(\omega_0)$ is the refractive index, which in general will depend upon the frequency of the wave, the phase velocity is

$$\begin{aligned} v_{\mathrm{ph}_0} &= \omega_0/k_0 \\ &= c/n(\omega_0) \end{aligned} \qquad [6.19]$$

Comparison with [6.18] shows that

$$A(\omega_0) = [1 - n(\omega_0)^2]\,\omega_0^2/c^2 \qquad [6.20]$$

Really we haven't shown very much here, merely that equations [6.18] and [6.19] are consistent. We can, however, make an interesting physical prediction if we make a transformation to a frame of reference in which the medium is moving. Suppose we go to a frame of reference that has a velocity $-\boldsymbol{v}$ with respect to the rest frame of medium. Then in that frame the medium will be moving with a velocity $\boldsymbol{v}$. If in this frame the frequency and wave number of the wave are ω and k, then using the inverse Lorentz transformation (equations [6.9]), putting $\omega' = \omega_0$ and $k' = k_0$, we get for the phase velocity in the new frame

$$v_{ph} = \omega/k = (\omega_0 + vk_0)/(k_0 + v\omega_0/c^2)$$
$$= (v_{ph_0} + v)/(1 + vv_{ph_0}/c^2) \qquad [6.21]$$

(Note that we could have obtained this result by using the inverse to the Lorentz transformation for the velocities (equations [4.8]).) If, as is almost always the case, $v \ll c$, then this expression can be simplified as follows

$$v_{ph} \triangleq (v_{ph_0} + v)(1 - vv_{ph_0}/c^2)$$
$$\triangleq c/n(\omega_0) + v(1 - 1/n(\omega_0)^2) \qquad [6.22]$$

This result incorporates the famous Fizeau 'drag factor', $(1 - 1/n^2)$, which in pre-Einsteinian relativity had to be explained by a dragging of the aether. In the correct theory it comes out quite naturally from the Lorentz transformation.

We now turn our attention to our second example, the propagation of waves in a metal. Here we have to be very careful; a metal does not transmit electromagnetic waves at all frequencies. The simple Drude theory of a metal gives the following criteria for the transmission of such waves. If ν is the frequency of collisions between electrons and ω_p is the so-called plasma frequency, defined as

$$\omega_p = (Ne^2/m\epsilon_0)^{1/2} \qquad [6.23]$$

where N is the number of electrons per unit volume and m and e are the mass and charge of the electron respectively, then electromagnetic waves propagate provided

$$\omega \ll \nu \qquad \omega > \omega_p \qquad [6.24]$$

If these criteria are satisfied then the dispersion relation for the waves passing through the metal is

$$\omega_0^2 - c^2 k_0^2 = \omega_p^2 \qquad [6.25]$$

Comparison with equation [6.16] immediately identifies A for this example as ω_p^2/c^2. Equation [6.18] then gives for the phase velocity

$$v_{ph_0} = c/(1 - \omega_p^2/\omega^2)^{1/2} \qquad [6.26]$$

which is always greater than c. The group velocity

$$= d\omega/dk$$

is obtained from equation [6.25]:

$$v_{gr} = c^2k/\omega = c[1-(\omega_p^2/\omega^2)]^{1/2} \qquad [6.27]$$

which is always less than c.

This then is as far as we shall take the discussion of the propagation of waves in a medium. We stress again that we have not in this chapter used the principle of relativity; what we have used is the transformation (the Lorentz transformation) that we derived from ensuring that the laws of physics should be form invariant. One of the consequences of our discussion is that, for frames which are in constant relative motion, we found the value of the phase was an invariant in going from one reference frame to another. Therefore, in going on in Chapter 7 to consider the interaction of electromagnetic waves with particles, we have only to consider how the amplitude of the wave transforms in going from one frame to another. However, before doing so we derive the Doppler effect.

6.5 The Doppler effect

Whether we are talking about waves propagating in free space or in a medium, the frequency (divided by c^2) and the three components of the wave vector form a four-vector. For simplicity consider the case where the wave is travelling in the x, y plane. Then

$$k_x = k\cos\phi \qquad k_y = k\sin\phi \qquad [6.28]$$

where ϕ is the angle between the x axis and the direction of propagation. Thus, using the first of these results and the last of equations [6.4], we can write for the frequency of the wave in the frame K'

$$\omega' = \gamma(\omega - vk\cos\phi) \qquad [6.29]$$

But

$$k = \omega/v_{ph}$$

hence we can write

$$\omega' = \omega[1-(v\cos\phi/v_{ph})]/[1-(v/c)^2]^{1/2} \qquad [6.30]$$

This then is the correct result for the Doppler effect. The numerator is the result one would get from the Galilean transformation, whilst the denominator results from the time-dilation effect in going from K to K'. The magnitude of the effect will depend on the particular waves we are considering, which will determine the magnitude of the phase velocity. Thus for electromagnetic waves in free space $v_{\mathrm{ph}} = c$, and hence

$$\omega' = \omega[1 - (v\cos\phi/c)]/[1 - (v/c)^2]^{1/2} \qquad [6.31]$$

This result should be compared with equation [5.24] where $v_0 = \omega'/2\pi$, $v = \omega/2\pi$. In particular if $\cos\phi = 1$, i.e. the wave is travelling in the x direction, then

$$\omega' = \omega[1 - (v/c)]^{1/2}/[1 + (v/c)]^{1/2} \qquad [6.32]$$

For $v/c \ll 1$ this reduces to the normal Doppler shift.

For waves travelling in a non-absorptive dielectric, the phase velocity is $c/n(\omega)$. Thus for the case $\cos\phi = 1$

$$\omega' = \omega[1 - (vn(\omega)/c)]/[1 - (v/c)^2]^{1/2} \qquad [6.33]$$

Finally, for electromagnetic waves travelling in a metal, the phase velocity is given by equation [6.26] and hence

$$\omega' = \omega\frac{[1 - (v/c)(1 - \omega_{\mathrm{p}}^2/\omega^2)]}{[1 - (v/c)^2]^{1/2}} \qquad [6.34]$$

The reader is invited to check for himself that the right-hand side can never be negative, since for waves to propagate $\omega > \omega_{\mathrm{p}}$. It is interesting to note that if $\omega = \omega_{\mathrm{p}}$, i.e. the waves just fail to propagate in the stationary medium, then the only effect that comes in is the time-dilation factor. In that situation the transformation equation for k, the third of equations [6.4], gives

$$k'_x = -v\omega/c^2$$

since from the dispersion relation (equation [6.26]) $k = 0$ at $\omega = \omega_{\mathrm{p}}$. Thus in the reference frame K' the wave actually propagates, but in the negative x direction.

We conclude this chapter by giving an example of the application of the Doppler effect.

6.6 Hubble's law

Edwin Hubble deduced from observations on the Doppler shift of the same spectroscopic line, but emitted from different and distant galaxies, that these galaxies were receding with a velocity which was proportional to their distance from our own galaxy. To show this we have to modify the result given in section 6.5. In the rest frame of our own galaxy, the Milky Way, let the recession velocity of the distant galaxy be v. The distant galaxy contains the source of radiation, say of frequency ω, and in its rest frame the Milky Way is receding with a velocity $-v$. Thus the frequency observed in our galaxy is given by equation [6.32] but with v replaced by $-v$. Hence

$$\omega' = \omega \left[\frac{1 + (v/c)}{1 - (v/c)}\right]^{1/2} \qquad [6.35]$$

The same result can be expressed in terms of the wavelength, since

$$\lambda = 2\pi c/\omega$$

i.e.

$$\lambda' = \lambda \left[\frac{1 - (v/c)}{1 + (v/c)}\right]^{1/2} \qquad [6.36]$$

Now λ is the wavelength of a particular line as measured when the source is at rest, i.e. on Earth. λ' is the wavelength of the same line emitted from the receding galaxy but measured on Earth. Experimental observation shows that these two wavelengths can differ by up to 25 per cent. For such a shift $\lambda'/\lambda = 1/1.25$, which gives, from equation [6.36], $v/c = 0.22$. The distant galaxies are obvious candidates for testing the Lorentz-FitzGerald contraction (see section 3.5) since it is a macroscopic body moving at a velocity comparable to the velocity of light. The only problem is to measure, say, the diameter of the galaxy in its own rest frame!

To return to Hubble's law, astronomical observation shows that the velocities as determined from the Doppler-shifted wavelengths are proportional to the distance of the galaxy from the Milky Way. This then is Hubble's law. It has the implication that, if we assume that the velocity of recession has remained constant since each galaxy was connected to the Milky Way, then the time that they have taken to reach their present positions is a constant, known as the Hubble time. Thus it is possible that all galaxies were merged together initially, i.e. there existed the possibility of a 'big bang'.

Chapter 7
Electrodynamics

7.1 Introduction

We are now in a position to bring together the two subjects that occupied the first half of this book, namely classical mechanics and electromagnetic theory. In doing so, we shall formulate the one outstanding issue that remains for us to discuss, namely the Lorentz invariant form of the equations of motion. In Chapter 2 we gave Newton's laws, albeit in a modified form, and these enabled us to write down unambiguously the equations of motion. Subsequently in Chapter 5 we showed that these equations were not form invariant under the Lorentz transformation. We avoided the issue of looking for the correct equations of motion by instead formulating the correct conservation equations. These latter equations, although useful for solving a number of problems, did not enable us to get a complete solution to a dynamic problem in the same way that the solution of Newton's equations of motion would have done.

We have already remarked on a number of occasions that there exists one fundamental difficulty in a correct relativistic formulation; namely, that because information cannot travel faster than the velocity of light, we cannot have the concept of instantaneous action at a distance. This is a concept that is central to the first of our reformulated laws of motion (equations [2.3] and [2.4]). In fact we have to be very careful; what follows from these equations was the equation of conservation of momentum (equation [2.7]). If we had initially assumed the law of conservation of momentum, then the law of motion (equations [2.3] and [2.4]) would have followed by differentiation with respect to time. Now in the correct relativistic formulation of the law of conservation of momentum and energy, the previous three-vector equation is replaced by a four-vector equation, the total energy-momentum four-vector.

We could equally well differentiate the four-vector equation with respect to time and obtain

$$\dot{\vec{\mathbf{p}}}_1 + \dot{\vec{\mathbf{p}}}_2 = 0$$

Again it is possible to interpret this equation as one particle instantaneously influencing the other, i.e. if $\mathrm{d}\vec{\mathbf{p}}/\mathrm{d}t$ changes for one particle it has to be instantaneously balanced by a change in the corresponding quantity for the other particle. Clearly this offends against the result that information cannot be transmitted instantaneously. The way out of this dilemma is obvious; it is not only that the sum of the two derivatives is equal to zero but that each derivative itself is zero. Hence not only is the sum of the two four-vectors equal to a constant but each vector itself is also a constant. We are led to the conclusion that the law of conservation of energy and momentum breaks down if there is any interaction between the particles. Actually this is not correct, for when the particles are interacting the momentum or energy 'missing' from the sum of the particle momenta or energies resides in the field responsible for the interaction. This, however, takes us too far ahead in our story; here we merely wish to emphasize that the Newtonian concept of action at a distance has to be abandoned.

The theory of electromagnetic phenomena, which led us into reconsidering the fundamental concepts of the relativistic transformation, also introduced a concept which we shall find useful in establishing the Lorentz invariant equations of motion. It will be remembered that in the theory of electromagnetism it was necessary to define electric and magnetic fields. These fields pervaded all of space, but their effect on the particle was purely local, i.e. the force on the particle due to the fields was determined only by the magnitude and direction of the field at the position of the particle. Furthermore, variations in the field travelled through space with the velocity of light, i.e. in the form of electromagnetic waves. Thus information from the source of the field, whether another particle or not, could only be conveyed at this velocity and certainly not instantaneously. There are several different fields in nature, e.g. gravitation, those responsible for the weak and strong interactions of elementary particle physics, etc. However, in order to be specific and for reasons of clarity, we shall discuss only the electromagnetic field.

7.2 Particle in an electromagnetic field

If we refer back to the beginning of Chapter 2 we find that there were three stages in formulating the Newtonian equations of motion. First, the recognition of the importance of acceleration; second, the definition of force as mass times acceleration; and finally, the existence of a law of force. When these arguments were applied to the case of a particle in an electromagnetic field, the force law was that given by Lorentz, i.e. equation [2.17]. This equation embodied both the definition of the electromagnetic field and a physical law (see the discussion following equation [2.17]). Thus in Newtonian mechanics the equation of motion is

$$m\mathbf{a} = e(\boldsymbol{\mathcal{E}} + \mathbf{u} \times \mathbf{B})$$

The question is: What is the correct, in the sense of being Lorentz as opposed to Galilean, invariant form of this equation?

There are two ways that we can proceed. We can either follow the method we used in postulating the law of conservation of momentum, i.e. we can try and guess the form of the equation of motion and in so doing ensure that it is form invariant under the Lorentz transformation, the test of its validity being the comparison of its predictions with experimental results. Alternatively, we can start from electromagnetic fields whose definitions ensure that we have the correct force at particle velocities tending to zero. We can then impose form invariance under the Lorentz transformation to find the equation of motion for any velocity. We shall adopt the latter approach on the grounds that we have reached the stage where we might reasonably be expected to accept the principle of relativity and the correctness of the Lorentz transformation. Further, such an approach has the advantage that we do not have to generate *ad hoc* arguments to justify the choice of one particular equation rather than another. The disadvantage is that we have to know how the fields transform from one inertial frame to another. However, since we now intend to use the principle of relativity as a tool rather than a test, we can use it to insist that Maxwell's equations are form invariant under the Lorentz transformation. Our first task will therefore be to insist on the invariance of Maxwell's equations to generate the equations for transforming the fields.

7.3 Transformation of the electromagnetic fields

We start from Faraday's law of induction in its differential form (equation [2.23]). In particular let us write down the z component

$$\frac{\partial \mathcal{E}_y}{\partial x} - \frac{\partial \mathcal{E}_x}{\partial y} = -\frac{\partial B_z}{\partial t}$$

Now in a similar manner that led to the derivation of equation [2.28], we can transform the partial derivatives in x, y and t to derivatives in x', y' and t' by using a similar procedure to that used for deriving equation [2.26] and [2.27] for the Galilean transformation. The results for the Lorentz transformation are (see exercise 1, Chapter 7, Appendix 3)

$$\frac{\partial}{\partial x} = \gamma\left[\frac{\partial}{\partial x'} - \frac{v}{c^2}\frac{\partial}{\partial t'}\right] \qquad \frac{\partial}{\partial y} = \frac{\partial}{\partial y'} \qquad \frac{\partial}{\partial z} = \frac{\partial}{\partial z'}$$

$$\frac{\partial}{\partial t} = \gamma\left[-v\frac{\partial}{\partial x'} + \frac{\partial}{\partial t'}\right] \tag{7.1}$$

Applying these results to Faraday's law above gives

$$\gamma\frac{\partial}{\partial x'}(\mathcal{E}_y - vB_z) - \frac{\partial \mathcal{E}_x}{\partial y'} = -\gamma\frac{\partial}{\partial t'}(B_z - v\mathcal{E}_y/c^2)$$

If Faraday's law of induction is to be form invariant under the Lorentz transformation, i.e. if the fields in the new frame are denoted by $\mathcal{E}_x'$, $\mathcal{E}_y'$ and B_z' then

$$\frac{\partial \mathcal{E}'_y}{\partial x'} - \frac{\partial \mathcal{E}'_x}{\partial y'} = -\frac{\partial B'_z}{\partial t'}$$

For the above two equations to be compatible we must have

$$\begin{aligned} \mathcal{E}'_y &= \gamma(\mathcal{E}_y - vB_z) \\ \mathcal{E}'_x &= \mathcal{E}_x \\ B'_z &= \gamma(B_z - v\mathcal{E}_y/c^2) \end{aligned} \qquad [7.2]$$

Similarly, by considering the y component of equation [2.24], we obtain the transformation equations for $\mathcal{E}_x$ (again), $\mathcal{E}_z$ and B_y, i.e.

$$\begin{aligned} \mathcal{E}'_z &= \gamma(\mathcal{E}_z + vB_y) \\ B'_y &= \gamma(B_y + v\mathcal{E}_z/c^2) \end{aligned} \qquad [7.3]$$

Finally the x component gives us

$$B'_x = B_x \qquad [7.4]$$

This last result is not as straightforward as the others. In order to prove it, we must assume that one other of Maxwell's equations (equation [2.16]) is form invariant (see exercise 2, Chapter 7, Appendix 3). Having obtained the transformation equations for the fields by imposing the invariance on equations [2.24] and [2.15], we can check our results by seeing if they leave the other two of Maxwell's equations (equations [2.14] and [2.22]) form invariant. The reader is encouraged therefore to do exercise 3, Chapter 7, in Appendix 3.

With equations [7.2], [7.3] and [7.4] enabling us to transform fields from one inertial frame to another, we are now in the position to derive the equation of motion of a charged particle in an electromagnetic field.

7.4 Equation of motion of a charged particle

As we remarked at the beginning of this chapter, the definitions of the electric and magnetic fields still stand if we make the proviso that the velocity of the particle tends to zero. Furthermore, we know that in the same limit the force is defined by mass times acceleration. Therefore if we move to the instantaneous rest frame of the particle, we know what the equation of motion must be, i.e. if we denote by primes quantities in the rest frame, we have

$$m\mathbf{a}' = e\boldsymbol{\mathcal{E}}' \qquad [7.5]$$

where m is the rest mass and e is the charge on the particle. As it is written, the left-hand side is not part of a four-vector, since equations [4.9] show $\mathbf{a}'$ does not transform like the Lorentz transformation. However, in this frame, and only in this frame,

$$m\vec{\mathbf{a}}' = \mathrm{d}\vec{\mathbf{p}}'/\mathrm{d}\tau$$

which is a four-vector. τ is the rest frame, or proper time, and an invariant under the Lorentz transformation $\vec{\mathbf{p}}$ is a four-vector and hence so is $\mathrm{d}\vec{\mathbf{p}}$. The procedure for discovering the correct equation of motion is to write down the spatial components of equation [7.5]. (We have a problem in interpreting the fourth component. This we will return to below.) These three equations are then transformed to the laboratory frame where the particle has a velocity u in the x direction, i.e. the x direction is chosen to be the direction of the particle velocity in the laboratory frame.

Consider first, for simplicity, the y spatial component. Under the Lorentz transformation $\mathrm{d}p_y' = \mathrm{d}p_y$, and using the first of equations [7.2], remembering that for this case $v = u$, we get

$$\frac{\mathrm{d}p_y}{\mathrm{d}\tau} = \gamma(\mathcal{E}_y - uB_z)$$

but since from equation [5.7]

$$\mathrm{d}t = \gamma \mathrm{d}\tau$$

this reduces to

$$\frac{\mathrm{d}p_y}{\mathrm{d}t} = (\mathcal{E}_y - uB_z) \qquad [7.6]$$

This then is the equation of motion for the y direction. It would be identical to the Newtonian equation with the Lorentz force if

$$\mathrm{d}p_y/\mathrm{d}t = ma_y$$

However that equality only holds in the rest frame. A similar argument gives for the z component

$$\frac{\mathrm{d}p_z}{\mathrm{d}t} = (\mathscr{E}_z + uB_y) \qquad [7.7]$$

When we consider the x component of the equation of motion we encounter a difficulty. This is because in transforming $\mathrm{d}p'_x$ to the laboratory frame we necessarily introduce the fourth component of the four-vector. Now $\mathrm{d}p'_x$ transforms in the same way as p'_x, i.e. as given by the first of equations [5.16]. Thus the equation of motion for the x direction is, in the laboratory frame,

$$\gamma\left[\frac{\mathrm{d}p_x}{\mathrm{d}\tau} - \frac{u}{c^2}\frac{\mathrm{d}E}{\mathrm{d}\tau}\right] = \gamma^2\left[\frac{\mathrm{d}p}{\mathrm{d}t} - \frac{u}{c^2}\frac{\mathrm{d}E}{\mathrm{d}t}\right] = e\mathscr{E}_x \qquad [7.8]$$

since from the second of equations [7.2] $\mathscr{E}_x = \mathscr{E}'_x$ and we have used equation [5.7]. The difficulty here is how to deal with the quantity ($\mathrm{d}E/\mathrm{d}t$). If we go back to the rest frame of the particle, then the equivalent quantity, i.e. the time-like component of ($\mathrm{d}\vec{\mathbf{p}}'/\mathrm{d}\tau$), is ($\mathrm{d}E'/\mathrm{d}\tau$), which is of course the rate of change of energy of the particle. Now remembering that in the rest frame of the particle we can use Newtonian dynamics, it follows therefore that the rate of change of energy is equal to the rate at which work is being done on the particle. Again, since the particle is at rest, this quantity will be equal to zero. If we transform this result to the laboratory frame, using the equivalent of the last of equations [5.16], we obtain

$$-\frac{u}{c^2}\frac{\mathrm{d}p_x}{\mathrm{d}\tau} + \frac{1}{c^2}\frac{\mathrm{d}E}{\mathrm{d}\tau} = 0$$

or

$$\frac{\mathrm{d}E}{\mathrm{d}t} = u\frac{\mathrm{d}p_x}{\mathrm{d}t} \qquad [7.9]$$

Substituting in equation [7.8] we obtain finally for the equation of motion in the x direction

$$\frac{\mathrm{d}p_x}{\mathrm{d}t} = e\mathscr{E}_x \qquad [7.10]$$

The equation of motion for the time-like component can be obtained from equation [7.9] using equation [7.10], i.e.

$$\frac{dE}{dt} = ue\mathscr{E}_x \qquad [7.11]$$

which again is to be interpreted as the rate at which the electric field is doing work on the particle. (The magnetic force is always at right angles to the velocity of the particle and therefore does no work.) Equations [7.6], [7.7] and [7.10] may be summarized in one three-vector equation:

$$\frac{d\mathbf{p}}{dt} = e(\boldsymbol{\mathscr{E}} + \mathbf{u} \times \mathbf{B}) \qquad [7.12]$$

remembering that the instantaneous direction of $\mathbf{u}$ was in the x direction. This equation is identical to the corresponding Newtonian equation of motion if the left-hand side of the Newtonian expression is written $(d\mathbf{p}/dt)$ rather than $m\mathbf{a}$. The difference between the two equations is that in Newtonian mechanics $(d\mathbf{p}/dt)$ is always equal to $m\mathbf{a}$, whereas in Einstein's formulation the equality holds only in the rest frame (where agreement between the two formulations must always hold).

Equations [7.11] and [7.12] complete our self-imposed task of finding the equations of motion of a particle in an electromagnetic field. From the argument that we gave to construct them, they must be form invariant under the Lorentz transformation. What remains therefore is to test their predictions against experiment. We do this by looking next at a number of simple experiments.

7.5 Charged particle in a constant electric field

Suppose that the electric field is in the y direction, i.e.

$$\boldsymbol{\mathscr{E}} = \hat{\mathbf{j}}\mathscr{E}$$

Then for this electric field and for zero magnetic field the components of equation [7.12] become

$$\frac{dp_x}{dt} = \frac{dp_z}{dt} = 0 \qquad \frac{dp_y}{dt} = e\mathscr{E} \qquad [7.13]$$

For reasons which will appear later we choose the following initial conditions

$$x = y = z = 0$$
$$u_x = u_0 \qquad u_y = u_z = 0 \qquad [7.14]$$

Equation [7.13] integrates immediately and, with the above initial conditions, gives

$$p_x = \frac{mu_0}{(1-(u_0/c)^2)^{1/2}} \qquad p_y = e\mathscr{E}t \qquad p_z = 0 \qquad [7.15]$$

Equations [5.9] express the components of **p** in terms of the components of the particle velocity **u**. Thus substituting in equations [7.15] we get

$$\frac{mu_x}{(1-(u/c)^2)^{1/2}} = \frac{mu_0}{(1-(u_0/c)^2)^{1/2}}$$
$$\frac{mu_y}{(1-(u/c)^2)^{1/2}} = e\mathscr{E}t \qquad [7.16]$$
$$u_z = 0$$

Remembering that

$$u_x^2 + u_y + u_z^2 = u^2$$

we can solve for u to give, after a little algebra,

$$u = \frac{u_0^2 + \beta^2 t^2 c^2}{1 + \beta^2 t^2} \qquad [7.17]$$

where

$$\beta = (1-(u_0/c)^2)^{1/2} e\mathscr{E}/mc \qquad [7.18]$$

If we substitute this result back into the equations of motion (equations [7.16]), all three equations can then be integrated using the initial conditions (equations [7.14]). Thus

$$x = u_0 \int_0^t \frac{1}{(1 + \beta^2 t'^2)^{1/2}} \, dt'$$

This integral is standard and can be expressed as an inverse hyperbolic sine function, i.e.

$$x = (u_0/\beta)\sinh^{-1}(\beta t) \qquad [7.19]$$

Before going on to solve for the other two coordinates, we should discuss this result. The original equations of motion had only a force in the y direction. In classical Newtonian mechanics the immediate result would be that the only acceleration would also be in the y direction. Equation [7.19] shows that this is not true for Lorentz invariant mechanics. In fact if we differentiate that equation we find

$$u_x = \frac{u_0}{(1 + \beta^2 t^2)^{1/2}}$$

which is certainly not a constant! Thus in insisting on Lorentz invariance we have to sacrifice the Newtonian result that force and acceleration act in the same direction. Only in the rest frame of the particle does that result still hold (where indeed the whole of Newtonian mechanics has to be valid).

In a similar manner to the derivation of the time path for the x coordinate, we find for the y coordinate

$$y = \beta c \int_0^t \frac{t'}{(1 + \beta^2 t'^2)^{1/2}}\, dt'$$

The integral is easily performed to give

$$y = (c/\beta)[(1 + \beta^2 t^2)^{1/2} - 1] \qquad [7.20]$$

Finally, and very easily using equation [7.16] and the initial conditions, equation [7.14], $z = 0$.

We can eliminate t from equations [7.19] and [7.20] to obtain the orbit of the particle:

$$y = (c/\beta)[\cosh(\beta x/u_0) - 1] \qquad [7.21]$$

In the classical case

$$y = e\mathcal{E}t^2/2m$$

$$x = u_0 t$$

Hence the orbit becomes

$$y = (e\mathscr{E}/2m)(x/u_0)^2 \qquad [7.22]$$

It is left as an exercise for the reader to show how, and under what circumstances, these two results become equivalent (see exercise 4, Chapter 7, Appendix 3).

7.6 Charged particle in a constant magnetic field

In Newtonian mechanics the answer to this problem is well known. The stable trajectories are circles, the radii of which are found by equating the force due to the magnetic field, which always acts at right angles to the field, and the velocity to the so-called 'centrifugal force', i.e.

$$mu^2/r = eBu$$

Hence

$$r = mu/eB \qquad [7.23]$$

Whether these circular orbits are consistent with the more general relativistic dynamics is a matter of some importance in the design of particle accelerators. At low velocities it is certainly possible to maintain a beam of particles in a circular orbit. The question is, if the particles are accelerated up to velocities close to the velocity of light, would they remain in the circular containing vessel?

Let the constant magnetic field be in the z direction and the initial conditions be the same as for the previous problem, i.e. given by equations [7.14]. For $\mathbf{B} = \hat{\mathbf{k}}B$ the equation of motion (equation [7.12]) becomes

$$\frac{\mathrm{d}\mathbf{p}}{\mathrm{d}t} = e(\hat{\mathbf{i}}u_yB - \hat{\mathbf{j}}u_xB) \qquad [7.24]$$

It follows immediately that $(\mathrm{d}p_z/\mathrm{d}t) = 0$ and thus, because of the choice of initial conditions, that $z(t) = 0$. Thus, as in the Newtonian solution, the motion is purely in the x, y plane. The x and y components of equation [7.24] can be integrated immediately and, using the first two equations of [5.9], which express the momenta in terms of the corresponding velocities, we obtain

$$\frac{m(\mathrm{d}x/\mathrm{d}t)}{(1-(u/c)^2)^{1/2}} = eBy + \frac{mu_0}{(1-(u_0/c)^2)^{1/2}}$$

$$\frac{m(\mathrm{d}y/\mathrm{d}t)}{(1-(u/c)^2)^{1/2}} = -eBx \qquad [7.25]$$

The equation of the trajectory can be found by dividing the second equation by the first:

$$\frac{\mathrm{d}y}{\mathrm{d}x} = \frac{-eBx}{eBy + mu_0/(1-(u_0/c)^2)^{1/2}}$$

The variables separate so the equation may be integrated to give

$$eBy^2/2 + mu_0 y/(1-(u_0/c)^2)^{1/2} = -eBx^2/2$$

This may be rearranged to give the final expression for the trajectory

$$(y-r)^2 + x^2 = r^2 \qquad [7.26]$$

This is of course a circle of radius

$$r = mu_0/eB(1-(u_0/c)^2)^{1/2} \qquad [7.27]$$

and centred at $y = r$. This result is identical to the Newtonian result with the exception that the Newtonian radius is to be divided by

$$(1-(u_0/c)^2)^{1/2}$$

Thus if $u_0/c = 0.9$ then the radius is increased, over the Newtonian one, by a factor of 2.3, a difference which no accelerator designer could afford to ignore!

Finally it is left for the reader (see exercise 5, Chapter 7, Appendix 3) to check that the actual time path of the particle is given by

$$x = r\sin(\omega t) \qquad y = r(1-\cos(\omega t)) \qquad [7.28]$$

where

$$\omega = u_0/r = eB(1-(u_0/c)^2)^{1/2}/m \qquad [7.29]$$

7.7 Constant electric and magnetic fields at right angles ($\mathcal{E} > cB$)

We now combine the two cases discussed above so that we have a constant electric field in the y direction and a constant magnetic field in the z direction. The straightforward way to solve this problem would be to write down the equation of motion (equation [7.12]) for these fields and then solve them. However there is an alternative procedure which will enable us to exploit the special solutions we have already obtained and to practise transforming from one reference frame to another.

The idea is very simple. We have seen that in going from one reference frame to another, moving with a constant velocity with respect to the first, magnetic and electric fields become mixed. In particular it is possible to have, say, in one inertial frame just an electric field. However when the same physical situation is considered from a different inertial frame, a measurement of the fields will indicate both magnetic and electric fields present.

To be particular, suppose in the K' frame there is only an electric field present in the y' direction. Then we have already solved that problem (at least for a particular set of initial conditions). The solution is given by equation [7.19] and [7.20] (placing primes on the space-time coordinates to comply with now being in the K' frame), i.e.

$$\begin{aligned} x' &= (u_0'/\beta')\sinh(\beta' t') \\ y' &= (c/\beta')((1+\beta'^2 t'^2)^{1/2} - 1) \end{aligned} \qquad [7.30]$$

where

$$\beta' = (1-(u_0'/c)^2)^{1/2} e\mathcal{E}'/mc \qquad [7.31]$$

and u_0' is the initial velocity in the x' direction in the K' frame.

As a first step we see what this solution looks like as viewed from the K, or laboratory, frame. To do this we simply have to make use of the inverse Lorentz transformation (equations [4.7]), i.e.

$$\begin{aligned} x &= \gamma(x' + vt') \\ y &= y' \\ \gamma &= 1/(1-(v/c)^2)^{1/2} \end{aligned} \qquad [7.32]$$

and v is the relative velocity of K' with respect to K. If we now substitute for x' and y' from equation [7.30] we get

$$\begin{aligned} x &= \gamma((u_0'/\beta')\sinh^{-1}(\beta' t') + vt') \\ y &= (c/\beta')[(1 + \beta'^2 t'^2)^{1/2} - 1] \end{aligned} \qquad [7.33]$$

These two equations express the coordinates x, y of the particle in the K frame as functions of the time in the K' frame. Thus to complete the transformation we need the last equation of the inverse transformation, i.e.

$$\begin{aligned} t &= \gamma(t' + vx'/c^2) \\ &= \gamma(t' + (vu_0'/\beta' c^2)\sinh^{-1}(\beta' t')) \end{aligned} \qquad [7.34]$$

where we have used the first of equations [7.30]. In principle this equation can be solved to find t' as a function of t and then substituted in equation [7.33] to find x and y as functions of t. In practice equation [7.34] does not possess an inverse, so the procedures cannot be carried through. Nevertheless it is possible to find the equation of the orbit.

The second of equations [7.33] can be inverted to give t' as a function of y, i.e.

$$t' = (1/\beta')((1 + \beta' y/c)^2 - 1)^{1/2}$$

and then substituted into the first equation. Thus, after a little algebra and introducing the dimensionless variables

$$X = x\beta'/\gamma v \qquad Y = y\beta'/c \qquad [7.35]$$

we get

$$X = (u_0'/v)\sinh^{-1}((Y + 1)^2 - 1)^{1/2} + ((Y + 1)^2 - 1)^{1/2} \qquad [7.36]$$

This then is the orbit as seen by an observer in the frame K. However, we have not completed the problem, for the orbit still contains the initial velocity in the x' direction as measured in the K' frame, i.e. u_0'. If we use equations [4.8] for transforming the velocities from one inertial frame to another, and if the initial velocities of the particle, as measured in the K frame, are

$$u_x = u_0 \qquad u_y = 0 \qquad u_z = 0 \qquad [7.37]$$

then

$$u_0' = \frac{(u_0 - v)}{1 - u_0 v/c^2} \qquad [7.38]$$

Substitution of this result into the equation for the orbit (equation [7.36]) means that we have only two other parameters to identify with quantities that can be measured in the K frame. The two parameters are $\mathcal{E}'$, the electric field in the K' frame, which appears in the parameter β' (equation [7.31]), and v, the relative velocity of the two frames. The obvious physical quantities in the K frame that these parameters should be linked with are the electric and magnetic fields measured in that frame.

Equations [7.2] to [7.4] give the transformations of the fields in the K frame to the fields in the K' frame. (In fact we need the inverse of these equations. These are easily obtained in the usual manner by interchanging the primed and unprimed variables and changing the sign of v.) For the case where in the K' frame the only field is an electric one in the y' direction, we get for the fields in the K frame

$$\begin{array}{lll} \mathcal{E}_x = 0 & \mathcal{E}_y = \gamma\mathcal{E}' & \mathcal{E}_z = 0 \\ B_x = 0 & B_y = 0 & B_z = \gamma v \mathcal{E}'/c^2 \end{array} \qquad [7.39]$$

The parameters v and $\mathcal{E}'$ are then given by

$$v/c = B_z c/\mathcal{E}_y \qquad \mathcal{E}'^2 = \mathcal{E}_y^2 - c^2 B_z^2 \qquad [7.40]$$

Equations [7.36], [7.38] and [7.40] are thus the solution to the problem of a charged particle subject to a constant electric field, $\mathcal{E}_y$, in the y direction and a constant magnetic field, B_z, in the z direction, which is initially at the origin and whose initial velocity is u_0 in the x direction. Figure 7.1 shows the orbits for various values of the parameter u_0'/v. It will be seen that whatever the value of this parameter the orbits are not bounded in the y direction. This is in direct contrast to the Newtonian result, where the orbits always have an upper bound to their y coordinate. In fact the classical orbits look like Fig. 7.2, which is actually drawn for the Lorentz invariant case when $\mathcal{E} < cB$ (see exercise 6, Chapter 7, Appendix 3).

The reason for this radically different behaviour is as follows. First, however, we must note that in fact we are dealing only with

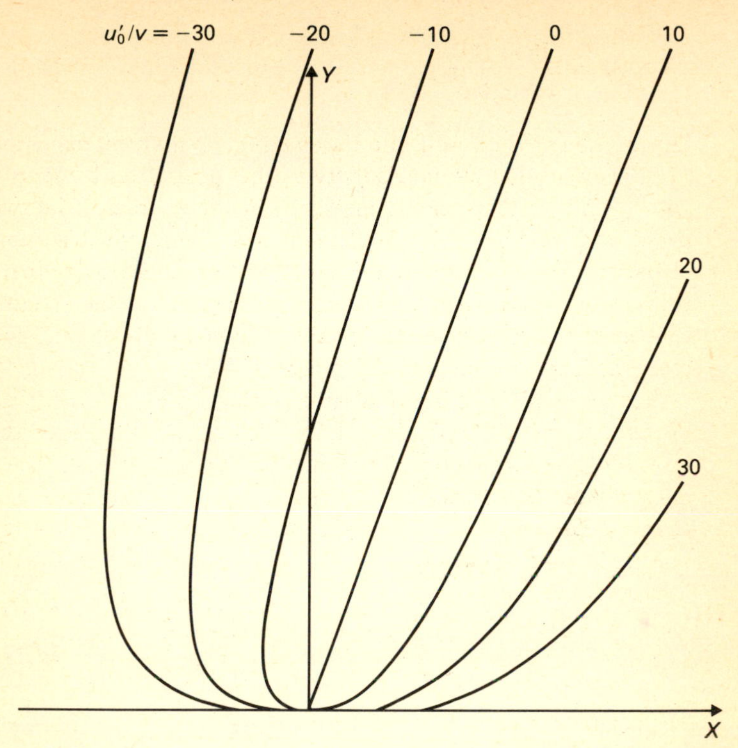

Fig. 7.1 The orbit of a charged particle when it is injected with a velocity along the x axis into an electric field $\mathscr{E}$ along the y axis and a magnetic field B along the z axis, for the case when $\mathscr{E} > cB$.

a range of relative field strengths. Since $v/c \leqslant 1$, it follows from the first of equations [7.30] that our solution is only valid in the range $\mathscr{E} \geqslant B$. The curvature of the orbits is due to the initial velocity u_0 and the magnetic force which acts perpendicularly to the particle velocity. Since u is increasing it might be expected that the magnetic force, which is proportional to u and has a component in the negative y direction, would eventually reduce the velocity in the y direction to zero and subsequently reverse its sign, thus putting an upper bound on the y coordinate of the orbit. However, the Lorentz invariant momentum contains the factor $1/(1-(u/c)^2)^{1/2}$, and because the electric field is stronger than the magnetic field (times c) the effective 'mass' of the particle is increasing faster than the effect of the magnetic force. In classical mechanics there is no such factor; the 'mass' is always the rest mass and the magnetic force is always able, eventually, to reduce the y component of the velocity to zero.

The obvious method of finding the solution when $\mathscr{E} < cB$ is to start from the case when the only field in the K' frame is a magnetic one. That is the subject of the next section.

7.8 Constant electric and magnetic fields at right angles ($\mathscr{E} < cB$)

We follow exactly the same procedure as in the previous section, but starting from the case that in the K' frame there is only a magnetic field in the z' direction and no electric field. Thus using the inverse Lorentz transformation and equations [7.28], the solution in the K' frame, we get the equivalent of equations [7.33], i.e.

$$\begin{aligned} x &= \gamma[r' \sin(\omega' t') + vt'] \\ y &= r'[\cos(\omega' t') - 1] \end{aligned} \qquad [7.41]$$

where

$$\begin{aligned} r' &= mu_0'/eB'(1-(u_0'/c)^2)^{1/2} \\ \omega' &= u_0'/r' \end{aligned} \qquad [7.42]$$

In this case it is impossible to write x as a function of y, or vice versa, so we regard equation [7.41] as the parametric equations of the orbit, t' being the parameter. If we introduce the dimensionless variables

$$X = \omega' x'/\gamma v \qquad Y = \omega' y/v \qquad \omega' t' = \tau' \qquad [7.43]$$

these parametric equations can be written

$$\begin{aligned} X &= (u_0'/v)\sin\tau' + \tau' \\ Y &= (u_0'/v)(\cos\tau' - 1) \end{aligned} \qquad [7.44]$$

The initial velocity in the frame K is related to u' in the same way as the previous case, i.e.

$$u_0' = \frac{(u_0 - v)}{1 - u_0 v/c} \qquad [7.45]$$

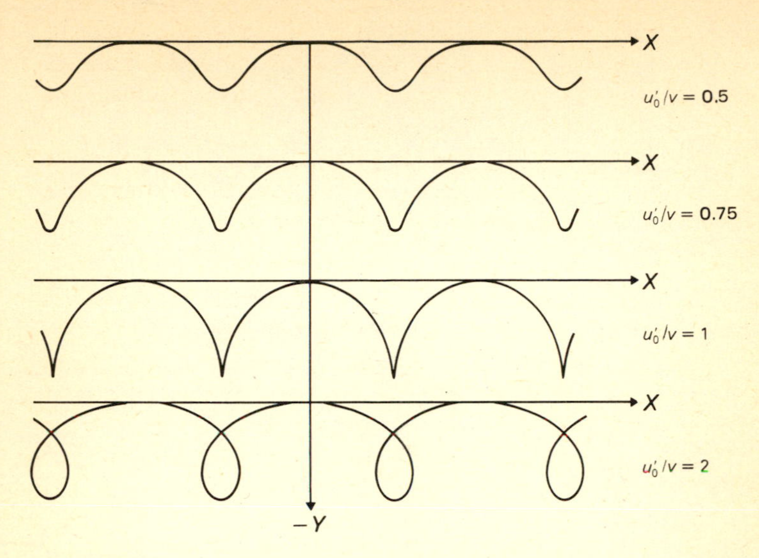

Fig. 7.2 The orbit of a charged particle when it is injected with a velocity along the x axis into an electric field $\mathcal{E}$ along the y axis and a magnetic field B along the z axis, for the case when $\mathcal{E} < cB$.

The relation between v and B' and the fields $\mathcal{E}_y$ and B_z, i.e. the equivalent of equations [7.40], is obtained from the inverse transformation to equation [7.2] to [7.4]:

$$\begin{aligned} v/c &= \mathcal{E}_y/cB_z \\ c^2B'^2 &= c^2B_z^2 - \mathcal{E}_y^2 \end{aligned} \qquad [7.46]$$

Thus equations [7.44] to [7.46] constitute a solution to the problem of a charged particle subjected to constant electric and magnetic fields which are at right angles and where $\mathcal{E} < cB$ (this condition follows from the first of equations [7.46]), the particle being initially injected at right angles to both fields. The orbits for various values of the parameter u_0'/v are shown in Fig. 7.2.

If u_0' is zero then Y is identically zero and $X = \tau$. The reason for this result is that if the initial velocity in the K' frame is zero, then the force on the particle is zero since the only field in that frame is a magnetic one. Thus in the frame K the particle must be moving with a constant velocity

$$v = \mathcal{E}_y/B_z \qquad [7.47]$$

where we have used equation [7.46]. Thus if we were to inject a number of particles with velocities lying in a narrow range and put a small aperture on the x axis, then only those particles with a velocity given by equation [7.47] would pass through the aperture (see exercise 7, Chapter 7, Appendix 3). This is the basis of the simple Wien filter.

7.9 Conclusion

This is as far as space will permit us to go. Even so we have covered an awful lot of physics from classical optics to the interaction of elementary particles. If we were to go on with the story, the next step would be to look at more complicated field configurations, followed by the interaction of a charged particle with time-varying fields. Then we should investigate whether it is possible to obtain a self-consistent solution to Maxwell's equations for the fields that interact with the charged particles, the subsequent motion of the particles being the currents that act as sources in Maxwell's equations. From a mathematician's point of view this set of equations does not possess a stable set of solutions; nevertheless a large and interesting amount of physics can be extracted. For a discussion of classical electrodynamics see Feynman (see Appendix 1).

The theory described in this book differs only from Newtonian mechanics for particles whose masses are extremely small, which in practice means the elementary particles. The behaviour and interaction of such particles is more properly described by quantum, rather than Newtonian, mechanics. To do this we would have to replace the Lorentz invariant equations of motion by a Lorentz invariant wave equation. Further, the Maxwell equations for the classical fields would have to be replaced by equations describing the quantized fields, or photons. Such a programme is far beyond the scope of this book but it is hoped that sufficient has been written to encourage the reader to pursue the subject further on his own. There is still a long way to go before the boundaries of the subject are reached; nevertheless it is a story which is as fascinating, if not more so, than the one told here.

Appendix 1
Bibliography

There is an enormous number of books on relativity, ranging from popular expositions to advanced monographs. Inevitably the choice given here reflects the author's preferences.

An excellent book which is very readable, full of puzzles and paradoxes with worked solutions, is E. F. Taylor and J. A. Wheeler, *Spacetime Physics*, W. H. Freeman, 1966. It is written for the U. S. undergraduate market and the mathematical level is very low; furthermore it restricts its discussion to the impact of relativity on mechanics.

A text closer to the content of this book is A. P. French, *Special Relativity*, Nelson, 1968, but again the discussion of electromagnetism is confined to the last chapter.

The classic R. P. Feynman, *Lectures on Physics*, Addison-Wesley, 1963, should be compulsory reading for anybody who has had a first course in the subject and wishes to get a deeper understanding.

Finally, mention should be made of the very readable and celebrated book, E. Whittaker, *A History of the Theories of Aether and Electricity*, Humanities Press, 1973. Although sometimes a little heavy going mathematically, it is nevertheless a fascinating account of the development of modern physics.

Appendix 2
Derivation of the Galilean and Lorentz transformations

In this appendix the assumption of linearity (equation [3.2]) is replaced by the more general assumption that

$$x' = X(x,t) \qquad t' = T(x,t) \qquad \text{[A2.1]}$$

where X and T are unknown functions to be determined. The equation analogous to equation [3.3] is

$$u'_x = \frac{u_x(\partial x'/\partial x) + (\partial x'/\partial t)}{u_x(\partial t'/\partial x) + (\partial t'/\partial t)} \qquad \text{[A2.2]}$$

The condition that if $u_x = v$, then $u'_x = 0$, gives

$$v\frac{\partial x'}{\partial x} + \frac{\partial x'}{\partial t} = 0 \qquad \text{[A2.3]}$$

and the condition that if $u'_x = -v$, then $u_x = 0$, gives

$$v\frac{\partial t'}{\partial t} + \frac{\partial x'}{\partial t} = 0 \qquad \text{[A2.4]}$$

Equation [A2.3] is a partial differential equation for the unknown function X which has the general solution

$$x' = X(x,t) = g(x - vt) \qquad \text{[A2.5]}$$

where g is an arbitrary function of its argument.

For the Galilean transformation $t' = t$ and hence [A2.4] gives

$$\frac{\partial x'}{\partial t} = -v \qquad \text{[A2.6]}$$

which combined with [A2.5] yields

$$g'(x - vt) = 1 \qquad [A2.7]$$

where g' denotes the derivative of the function with respect to its argument. Hence with the given initial conditions this may be integrated to give the Galilean transformation.

For the Lorentz transformation we need the equation for transforming the velocities in the y direction. From [A2.1] we get

$$u'_y = \frac{u_y}{u_x(\partial t'/\partial x) + (\partial t'/\partial t)} \qquad [A2.8]$$

Consider the case where in the K frame we have a ray of light propagating in the y direction, i.e. $u_x = 0$, $u_y = c$. Using the constancy of the velocity of light, we get

$$c^2 = u'_x + u'_y = v^2 + c^2/(\partial t'/\partial t)^2$$

where we have used equation [A2.8] and the result that if $u_x = 0$, then $u'_x = -v$.

Solving for the partial derivative gives

$$\frac{\partial t'}{\partial t} = \gamma = \frac{1}{[1 - (v/c)^2]^{1/2}} \qquad [A2.9]$$

Now from equations [A2.4] and [A2.5]

$$\frac{\partial t'}{\partial t} = -\frac{1}{v}\frac{\partial x'}{\partial t} = g'(x - vt) \qquad [A2.10]$$

Combining equations [A2.9] and [A2.10] and integrating using the given initial conditions gives

$$x' = g = \gamma(x - vt) \qquad [A2.11]$$

Next consider the case when $u_y = 0$ and $u_x = c$ and consequently $u'_y = 0$. Using equations [A2.2], [A2.9] and [A2.11], we get

$$u'_x = \frac{\gamma c - \gamma v}{c(\partial t'/\partial x) + \gamma}$$

Since u' must be equal to c

$$\frac{\partial t'}{\partial x} = -\gamma \frac{v}{c^2} \qquad \text{[A2.12]}$$

With the usual initial conditions the Lorentz transformation is obtained by integrating equations [A2.10] and [A2.12].

Appendix 3
Exercises

Chapter 1

1 Suppose a rod of length l is lying in the yz plane of a Cartesian coordinate system. A new coordinate system has the same x axis as the first but its y' and z' axes are rotated by an angle ϕ with respect to the corresponding axes of the first system. Show that in the new system the length of the rod has the same value.

2 $\mathbf{r}$ and $\mathbf{v}$ are the position and velocity of an isolated particle and $\mathbf{r}_1$ and $\mathbf{v}_1$ are constant vectors. In different reference frames the motion of the particle is described by the following formulae:
(a) $\mathbf{v} = \mathbf{v}_1$
(b) $\mathbf{r} = \mathbf{v}_1 t$
(c) $\mathbf{r} = \mathbf{v}t + \mathbf{r}_1$
(d) $d\mathbf{v}/dt = \mathbf{v}_1/a$ (a = constant scalar)
(e) $d\mathbf{v}/dt = \mathbf{g}$ (the acceleration due to gravity)
Decide which of the reference frames are inertial.

3 A particle moves with speed u at an angle ϕ to the x axis of a frame K. Show that in the frame K', moving at speed v parallel to the x axis of K, the particle moves at an angle ϕ' with the x' axis where ϕ' is given by

$$\tan \phi' = \frac{u \sin \phi}{u \cos \phi - v}$$

Chapter 2

1 Show for N particles that, if momentum is conserved, the centre

of mass, defined by

$$\mathbf{R} = (m_1\mathbf{r}_1 + m_2\mathbf{r}_2 + \ldots)/(m_1 + m_2 + \ldots)$$

has the equation of motion

$$\frac{d\mathbf{R}}{dt} = \text{constant}$$

This implies that the gross motion of the system can be treated as though it were a single particle. What will be the consequent motion of the centre of mass? The reference frame has to be inertial for Newton's laws to be valid. Therefore how could the definition of an inertial frame be widened to include the more realistic case, when there is more than one particle present?

2 Apply the coordinate transformation described in exercise 1, Chapter 1, to the laws of mechanics. Are they form invariant under this transformation?

3 Apply the Galilean transformation to the law of conservation of momentum and to the linear motion of the centre of mass. Are the resulting equations form invariant?

4 The equation of motion of a particle of charge q in an electromagnetic field is

$$m\frac{d\mathbf{u}}{dt} = q(\boldsymbol{\mathcal{E}} + \mathbf{u} \times \mathbf{B})$$

where m is the mass of the particle and $\mathbf{u}$ its velocity. Find the motion of the particle in the special cases:

(a) $\boldsymbol{\mathcal{E}} = \hat{\mathbf{j}}\mathcal{E}_0$ $B = 0$ $\mathcal{E}_0$ is a constant

(b) $\mathcal{E} = 0$ $\mathbf{B} = \hat{\mathbf{k}}B_0$ B_0 is a constant

In both cases the initial conditions are

$$x = y = z = 0 \quad u_x = u_0 \quad u_y = u_z = 0$$

where u_0 is a constant.

5 By choosing the axes of the reference frame so that the velocity of a charged particle is first along the y axis, then along the z axis, prove equations [2.20].

6 Complete the proof of the equations [2.27].

7 Prove by elementary arguments the formulae for the Doppler effect. How many independent cases do you have to consider?

Chapter 3

1 The (x,t) coordinates of three events A, B and C in a particular inertial frame are $(1,2/c)$, $(4,7/c)$, and $(6,5/c)$. (a) Which of the intervals between the two events are time-like and which are space-like? (b) What is the proper time (or proper distance) between the events? (c) Which event, if any, could have caused another event?

2 A particle enters our galaxy with a velocity of $0.5c$. How long would it take to traverse the galaxy in the rest frame of the particle and in the rest frame of the galaxy? (The galaxy has a diameter of the order of 10^5 light years. 1 light year = 9.5×10^{12} km.)

3 In Chapter 5 it is shown that the total energy E of a single particle of mass m is related correctly to its velocity u by

$$E = mc^2/[1-(u/c)^2]^{1/2}$$

For an electron $mc^2 = 0.5$ MeV. Show that if its total energy is 1.3 MeV then its velocity is $12c/13$. Determine the time it takes to travel a distance of 360 m and the time that has elapsed in its own rest frame.

4 The distance to a certain star is (in the Earth's coordinates) 20 light years. Calculate the speed relative to the Earth of a spaceship which travels to the star in 5 years (according to clocks on the spaceship).

5 A particle moving at a constant velocity traverses two counters fixed 2 m apart in the laboratory. As observed in the laboratory the second counter triggers 10^{-8} s after the first. What is the time difference between these two events with respect to the rest frame of the particle?

6 Draw the equivalent of Fig. 3.4 but for only one space dimension. Draw in the lines which separate the space-time continuum into regions where the interval is time-like and where it is space-like. On the same diagram plot the trajectory of a particle whose position is given by $x = ut$, where $u < c$ is a constant. Such a trajectory is known as the world line of the particle. Draw the world line when $x = at^2$, where a is a constant. Can this latter formula be correct for all times?

7 One of the most famous paradoxes arising from Einstein's theory is the so-called twins paradox. The twins are separated on their twenty-first birthday. One stays on Earth and the other goes on a spaceship journey into the galaxy which, according to his twin, lasts 50 years. Assuming that the travelling twin's journey is an outward journey at a uniform velocity of $0.7c$ lasting 25 years Earth time, followed by a return journey of the same length and velocity, and that the accelerations and decelerations have a negligible effect on the clocks, what is the age of the twin on Earth and what is the age of the travelling twin? The paradox which generated some controversy in the nineteen-fifties is this; is the twin really younger than his brother/sister, i.e. would he have the appearance of a younger man and a greater life expectancy? What do you think?

Chapter 4

1 Derive the Lorentz transformation using the assumptions of section 4.2, except that the condition that the velocity of light is an invariant is replaced by the condition that the interval is an invariant.

2 By using the fact that $a'_x = \mathrm{d}u'_x/\mathrm{d}t'$, $a'_y = \mathrm{d}u'_y/\mathrm{d}t'$, etc., derive equations [4.9] for the transformation of the accelerations under the Lorentz transformation.

3 Two spaceships A and B set off simultaneously from the Earth directly towards a planet which is at rest with respect to the Earth and at a distance of 100 light years. They travel at uniform speeds of $0.8c$ and $0.9c$ respectively. Spaceship A, on arrival, immediately sends a radio signal to B telling it its time of arrival and awaits ship B. (a) Draw the world lines for A and B (in the Earth's reference frame) illustrating this sequence of events. (b) What times are registered on the clocks of the two spaceships when they meet on the planet? (c) At what time, by its own clock, does B receive the radio signal from A? The clocks on both ships read zero when they leave Earth.

4 A particle moves with a speed u at an angle ϕ to the x axis of the frame K. Show that in frame K' the particle moves at an angle ϕ' to the x' axis, where ϕ' is given by

$$\tan \phi' = \frac{u \sin \phi}{\gamma(u \cos \phi - v)}$$

5 Rocket A is launched from Earth at a speed of $0.6c$. One hour later (as shown on the Earth's clock) rocket B is launched in the same direction and the two rockets pass each other at a time registered by A as 3.2 hours. (a) Find the time from launching and distance from the Earth of the event when the rockets pass as measured in a space-time frame attached to the Earth. (b) Find the launch velocity of rocket B and the time measured by B when it passes A. (c) A radio signal is emitted by A when B passes it. At what time (as measured by the clock on Earth) is this signal received on Earth?

6 A spaceship (frame K') travels away from Earth (frame K) at a speed $v = 3c/5$ along common x, x' axes. In the coordinate frame of the spaceship an explosion (event A) occurs at $x'_A = 9 \times 10^8$ m, $y'_A = 12 \times 10^8$ m, $z'_A = 0$, $t'_A = 3$ s. Light from the explosion reaches the spaceship (event B) and one second later (as measured in K') a radio signal is transmitted from the spaceship (event C). (a) Calculate the space-time coordinates of event A in the frame K. (b) Calculate the space-time coordinates of event B in the frame K'. (c) Show that the radio signal reaches Earth at a time $t = 18$ s Earth time. (The Earth and the spaceship are at the origins of K and K' which coincide at $t = t' = 0$.)

7 A spaceship at rest in the frame K' has a proper length of 100 m and is travelling along the x axis of the frame K with a speed of $v = 0.6c$. The tail of the spaceship passes the origin of K at the time $t = t' = 0$. (a) Calculate the time in K and K' at which the nose of the spaceship passes the point $x = 200$ m (event 1). At the same time as event 1 is measured in the K' frame, a bullet is fired from the tail (event 2) towards the nose at a speed (in K') of $u' = 0.5c$. (b) Show that the bullet strikes the nose (event 3) at the time (in K') $t'_3 = 10^{-6}$ s. (c) Calculate the space-time coordinates x_z and t_z (in K) of event 3.

8 In the K frame a source produces a cone of light symmetrically directed along the x axis, the semi-angle of the cone being ϕ. Describe what an observer sees who is at the origin of K'.

9 A triangle has vertices whose coordinates in the K frame are

(0,0), (2,1) and (1,2). Calculate the area of the triangle. What is the area of the triangle in the K' frame?

10 In the reference frame K' a circle is enscribed about the origin. What does an observer in the frame K see?

11

Planet	Distance from the Sun (10^6 km)	Period round the Sun	Diameter (km)
Mercury	58	88 days	5 000
Earth	150	365 days	12 800
Mars	228	687 days	6 780
Jupiter	778	12 years	142 600
Saturn	1 426	29 years	119 000
Uranus	2 868	84 years	51 500
Neptune	4 494	165 years	49 900

Would any of the above planets be, in principle, a better candidate than Venus for the proposed experiment described in section 4.3?

Chapter 5

1 Using the definition of a scalar product of two four-vectors (equation [5.3]), show that it has the same value in all inertial frames.

2 A neutron at rest beta-decays into a proton, an electron and a neutrino. Find the energy of the neutrino: (a) when the proton remains at rest; (b) when the electron remains at rest. The rest energies are, for the neutron 939.5 MeV, for the proton 938.2 MeV, for the electron 0.5 MeV, and the neutrino mass is approximately zero. Until comparatively recently the mass of the neutrino was taken to be identically zero; it is now thought to have a mass of around 35 eV. What difference, if any, would this make to the calculation above?

3 A pion moves with a constant velocity such that its kinetic energy is equal to its rest energy. It decays into two photons, which in the rest frame of the pion are at right angles to its original direction of motion. What are the energies and directions of the photons in the original frame of reference?

4 Generalize the calcuation of Compton scattering, given in section 5.10, to the case where the target electron is moving with a constant velocity, either by using the method described in Chapter 5 or by transforming equation [5.30] to an appropriate inertial frame.

5 A particle of mass m travelling at a speed u collides with a stationary particle of equal mass and they combine to form a new particle. Calculate the speed v and the mass M of the new particle, expressing your result in terms of m, u and $\gamma = 1/(1-(u/c)^2)^{1/2}$. The new particle subsequently breaks up into two particles, each of mass αm. Show that the momentum of each particle in the zero-momentum frame is given by

$$p = mc[(1+\gamma-2\alpha^2)/2]^{1/2}$$

and hence deduce a maximum value for α.

6 A beam of negatively charged pions, of variable energy E, hits a target of protons at rest. Neutrons and photons are produced according to the reaction

$$\pi^- + \mathrm{p} \rightarrow \mathrm{n} + \gamma$$

If the neutron emerges at 90° to the direction of the incoming pion, show that its energy is

$$E_\mathrm{n} = m_\mathrm{p}c^2 + \frac{m_\pi^2 c^2}{2(E + m_\mathrm{p}c^2)}$$

(The neutron and proton masses may be taken to be the same.)

7 The collision of high-energy pions with protons that are stationary in the laboratory frame may produce K and Σ mesons according to the reaction

$$\pi + \mathrm{p} \rightarrow \mathrm{K} + \Sigma$$

Show that the threshold energy for the pion in this reaction is

$$E_\pi^0 = \left[\frac{(m_\Sigma + m_\mathrm{K})^2 - m_\pi^2 - m_\mathrm{p}^2}{2m_\mathrm{p}}\right] c^2$$

where m_Σ, m_K, m_π and m_p are the rest masses of the Σ, K and π meson and the proton respectively.

8 A photon γ of energy E is incident upon a proton p at rest in the laboratory and leads to the production of a pion π by the reaction

$$\gamma + \mathrm{p} \rightarrow \pi + \mathrm{p}$$

Show that the threshold value of E is given by

$$E_0 = m_\pi c^2(1 + m_\pi/2m_\mathrm{p})$$

If the pion is emitted at 90° to the direction of the incident photon, show that its energy E is given by

$$E_\pi = \left[\frac{(Em_\mathrm{p} + m_\pi^2 c^2/2)}{E + m_\mathrm{p}c^2}\right] c^2$$

9 It is proposed to construct a proton accelerator so that anti-protons, i.e. particles of the same mass as the proton but of the opposite charge, can be produced. The reaction envisaged is

$$p + \mathrm{A} \rightarrow \mathrm{p} + \mathrm{A} + \mathrm{p} + \overline{\mathrm{p}}$$

where A is the target nucleus (mass M) and $\overline{\mathrm{p}}$ is the anti-proton. What is the minimum incident proton energy for which the accelerator should be designed? Would it be preferable to use carbon or uranium as the target? ($M_{\mathrm{p}}c^2 = 940$ MeV where M_{p} is the mass of the proton.)

Chapter 6

1 An astronaut possesses a medium-wave receiver and wishes to listen to a broadcast at 100 MHz. He has a spaceship at his disposal. How fast must he travel and should he fly towards or away from the transmitter?

2 The sun rotates once in about 24.7 days. The radius of the sun is about 7.0×10^8 m. Calculate the Doppler shift that we should observe for light of wavelength 500 nm from the edge of the Sun's disc near the equator. Is this shift towards the red end or the blue end of the spectrum?

3 Our galaxy rotates once in about every 200 million years and our Sun is located 30 000 light years from the galactic centre. Estimate the order of magnitude of the Doppler shifts, produced by this rotational motion, of visible light received by us from other galaxies. Explain how such estimates might be used to determine the speed of rotation.

4 A monochromatic beam of X-rays of wavelength λ are shone onto a gas of free electrons of mass m which are effectively at rest. The scattered X-rays are observed in a spectrometer at an angle θ. Derive an expression for their wavelength. If the same shift in wavelength were to occur not by scattering but by a moving detector, how fast would it have to move for the case where the incident photon energy was $0.5mc^2$ and $\theta = \pi/3$?

5 Derive the Fizeau drag coefficient by using the Lorentz transformation of the velocities. What would be the result if the Galilean transformation had been used instead?

6 Derive a wave equation for an electromagnetic wave travelling in a metal from the dispersion relation

$$\omega^2 - c^2k^2 = \omega_p^2$$

(Use the derivation of equation [6.15] as a guide.) Is this wave equation consistent with Maxwell's equations?

7 Take the point of view that electromagnetic radiation consists of a stream of particles with the energy-momentum relation $E = cp$. Hence derive the Doppler shift by transforming the energy and momentum to a different inertial frame and using the Planck relation between energy and frequency.

Chapter 7

1 Follow the procedure used in deriving the equations of the Galilean transformation for the differential operators in equations [2.26] and [2.27], using instead the Lorentz transformation (equation [4.6]) and its inverse (equation [4.7]), to derive equations [7.1].

2 Use the procedure illustrated in section 7.3 to derive the equations relating electric and magnetic fields in different inertial frames (equations [7.2] to [7.4]).

3 Show that with the relations derived in problems 2 and 3 above, the remaining two of Maxwell's equations are form invariant under the Lorentz transformation.

4 Under what condition would you expect the orbit (equation [7.21]) of a particle in an electric field to reduce to the classical orbit (equation [7.22])? Demonstrate that under this condition the reduction actually occurs.

5 Show that for a charged particle in a magnetic field solely along the z axis, the time development of the x and y coordinates is given by equations [7.28].

6 Write down Newton's equations of motion for a charged particle in a constant magnetic field in the z direction and a constant electric field in the y direction. Show that the solution has the same form, whatever the relative strengths of the fields.

7 It is required to filter out from a beam of electrons those with a specific velocity. Explain how this can be done using crossed electric and magnetic fields, including in your answer the necessary

formulae. If the spread in the velocities is initially 25 per cent about the mean velocity, what would be the spread in the final filtered velocities?

8 Solve the equations of motion of a charged particle injected with a constant velocity at right angles to mutually perpendicular constant electric and magnetic fields, for the special case where $\mathcal{E} = cB$. What makes this case special?

Appendix 4
Answers to exercises

Chapter 1

2 All but (d) and (e) are inertial.

Chapter 2

4 $y = q\mathcal{E}_0 t^2/2m, x = u_0 t, z = 0; r = (x^2 + y^2)^{1/2} = mu_0/qB.$

Chapter 3

1 AB is time-like, AC space-like and BC light-like; $\Delta\tau_{AB} = 4/c$, $\Delta\sigma_{AC} = 4$, $\Delta\tau_{CB} = 0$; A could have caused B, and C could have caused B.

2 1.7×10^5 years; 2×10^5 years.

3 1.3×10^{-6} s, 5×10^{-6} s.

4 $0.95c$.

5 7.4×10^{-9} s.

7 71; 56.7.

Chapter 4

3 62 years, 75 years; 70 years.

5 4 hours; $0.8c$, 2.8 hours; 6.4 hours.

6 (18×10^8 m, 12×10^8 m, 0, 6 s); (0, 0, 0, 9s).

7 $200/c$, $100/c$; (225 m, $325/c$).

9 $3/2$; $3/2\sqrt{1-(v/c)^2}$.

Chapter 5

2 0.55 MeV; 0.80 MeV.

3 $E = \gamma mc^2/2$; $\pi/6$.

4 $$E = \frac{E_{\gamma_1}(E_{e_1} - p_{e_1} c \cos\phi)}{E_{e_1} + E_{\gamma_1} - cp_{e_1} \cos\theta}$$

where θ is the angle between the two photons.

5 $[(1+\gamma)/2]^{1/2}$.

9 Threshold kinetic energy $= 2M_p c^2 [1 + 2M_p/M_A]$.

Chapter 6

1 $(c - v)/c = 2 \times 10^{-4}$; away.

2 $\Delta\lambda = 0.0035$A; Blue when approaching the Earth and red when receding.

3 $\Delta\lambda/\lambda = 9.4 \times 10^{-4}$.

4 $v/c = 9/41$.

Index

STUDENT PHYSICS SERIES

Series Editor:
Professor R. J. Blin-Stoyle, FRS
Professor of Theoretical Physics, University of Sussex

The aim of the *Student Physics Series* is to cover the material required for a first degree course in physics in a series of concise, clear and readable texts. Each volume will cover one of the usual sections of the physics degree course and will concentrate on covering the essential features of the subject. The texts will thus provide a core course in physics that all students should be expected to acquire, and to which more advanced work can be related according to ability. By concentrating on the essentials, the texts should also allow a valuable perspective and accessibility not normally attainable through the more usual textbooks.

CLASSICAL MECHANICS

A course in classical mechanics is an essential requirement of any first degree course in physics. In this volume Dr Brian Cowan provides a clear, concise and self-contained introduction to the subject and covers all the material needed by a student taking such a course. The author treats the material from a modern viewpoint, culminating in a final chapter showing how the Lagrangian and Hamiltonian formulations lend themselves particularly well to the more 'modern' areas of physics such as quantum mechanics. Worked examples are included in the text and there are exercises, with answers, for the student.

B. P. Cowan

Dr Brian Cowan is in the Department of Physics, Bedford College, University of London

ISBN 0-7102-0280-6
About 128 pp., diagrams, 129 mm x 198 mm, April 1984

STUDENT PHYSICS SERIES

ELECTRICITY AND MAGNETISM

Electromagnetism is basic to our understanding of the properties of matter and yet is often regarded a difficult part of a first degree course. In this book Professor Dobbs provides a concise and elegant account of the subject, covering all the material required by a student taking such a course. Although concentrating on the essentials of the subject, interesting applications are discussed in the text. Vector operators are introduced at the appropriate points and exercises, with answers, are included for the student.

E. R. Dobbs

Professor Roland Dobbs is Hildred Carlile Professor of Physics at the University of London.

ISBN 0-7102-0157-5
About 128 pp., 198 mm x 129 mm, diagrams, April 1984